'[The] writing is noticeably more stylish than
normally found in popular science books . . .
an engaging and well-researched book.'

FORTEAN TIMES

'You already thought Mars was a pretty weird
place – this excellent book will show you that it's far
weirder than you could ever have imagined.'

MAT COWARD, AUTHOR OF *ACTS OF DESTRUCTION*

'A complete history of the red planet . . . layer by layer, from
Mars' nascent stages as "grains of dust and starlight" through
to the formation of its crust and planet-shattering meteor
strikes, Morden unravels the planet's story and trajectory.
The author's background as both a planetary geologist and a
science-fiction author gives this story of Mars its unique spin.'

REACTION

'Superb . . . full of joy and wonder . . . brilliant.'

SWANSEA BAY MAGAZINE

'Brought the storytelling skills and narrative drive
of science fiction to telling the story of Mars.'

POPULARSCIENCE.COM

'With an engaging tone, Modern takes the reader
across the planet's surface and back and forward in
time on Mars with his evocative perspectives.'

SPACEFLIGHT MAGAZINE

THE
RED
PLANET

THE
RED
PLANET

A NATURAL HISTORY
OF MARS

SIMON MORDEN

Elliott&Thompson

First published 2021 by
Elliott and Thompson Limited
2 John Street
London WC1N 2ES
www.eandtbooks.com

This paperback edition first published in 2022

ISBN: 978-1-78396-661-5

9 8 7 6 5 4 3 2 1

A catalogue record for this book is available from
the British Library.

Cover image: NASA/JPL-Caltech
Typesetting: Marie Doherty
Printed by CPI Group (UK) Ltd, Croydon, CR0 4YY

For my mum, and in memory of my dad,
neither of whom ever told me I needed to get a proper job

CONTENTS

Enter, Stage Left   xiii

PART ONE: EXPLAINING MARS
Dawn on Mars   3
Mars as an Unreliable Narrator   9
Why Is Mars So Different?   13
Mapping Mars   17

PART TWO: BEFORE THE BEGINNING
The Giant Molecular Cloud   25
Accretion   31
Planetary Embryos   35

PART THREE: EARLY MARS
Mars at the Start   43
Planetary Melting   47
Crater Formation   53
Crater Counting   57
Martian Meteorites   63
The Great Dichotomy   67
The Great Dichotomy Convection Theory   71
The Great Dichotomy Impact Theory   77
Phobos and Deimos   83
The Early Martian Atmosphere   87

PART FOUR: THE NOACHIAN

Sailing on an Endless Sea   93

The Hellas Impact   97

The Start of the Noachian   103

Obliquity and Eccentricity   107

Introducing Tharsis   111

We Need to Talk about Water   115

The Northern Ocean   119

Life   123

Tharsis Rises   131

Lake Eridania   135

PART FIVE: THE HESPERIAN

The Hesperian Climate Change   143

The Beginning of the Cryosphere   147

Valles and Chaoses   151

Here Are Giants   157

Valles Marineris   161

Olympus Mons   169

Elysium   175

The Medusae Fossae Formation   179

True Polar Wander   185

The Ice Caps   191

PART SIX: THE AMAZONIAN

A World of Ice   197

Into the Amazonian   201

The Amazonian Climate   205

Equatorial Ice   209

High Latitude Ice   215

The Polar Regions   221
The Dust Cycle   227
Amazonian Volcanism   233

PART SEVEN: THE FUTURE

What Can We Make of Mars?   239
We Are the Martians   247

Acknowledgements   251
Bibliography   253
Index   265
About the Author   274

ENTER, STAGE LEFT

I grew up in a village with no street lights. The night sky was a vast bowl of dark, its edges lightened by the sodium glare of the nearby towns, its interior scratched by the navigation lights of airliners drifting down towards Heathrow. Venus, the bright morning and evening star, could hang on the horizon like a white jewel for days on end, and Jupiter and Saturn danced high in the sky. They wandered through the constellations, but were otherwise indistinguishable from stars. Mars, though, was different: from Earth it was clearly visible as a wavering point of red light. Identifiably, unmistakably red.

Mars was an inspiration to me, even if it always remained tantalisingly out of reach. But I did have a close encounter with it. I had progressed in an entirely unexceptional way from a degree in geology to a master's in geophysics, and a mid-course sidestep into a doctorate on the magnetic properties of meteorites. The secrets of how and when the planets formed are locked up in these rocks, and the information I needed was held in microscopic fragments of raw, unaltered iron–nickel alloys. Later, in my first and only postdoctoral post, I ordered a piece of a meteorite from NASA's collection. It duly arrived and I ran the usual tests on it, only to be disappointed. The magnetic minerals had been weathered to rust. It was useless to my research, so I put it to one side and concentrated on other, more promising samples.

I had been right that there was no trace of iron metal in that sample; it had all been converted to iron oxides. But it hadn't happened here on Earth, and I should have looked harder, questioned more. Six months later, a laboratory in Japan analysed the trapped gases in the rock and announced their results. The previously miscategorised ALH84001 had formed on Mars, and the few grams of material I'd already sent back to the USA were now part of a tiny group of known Martian rocks, far too precious for a junior researcher to work on. I'd missed my chance. The economic tide receded. I left academic research. Eventually, I became a science-fiction writer. I even wrote stories set on Mars.

That lost opportunity was thirty years ago, always remembered with a sense of regret. Then, this unlooked-for grace. I'm fully aware of how ridiculous it sounds, but writing this book feels like completion, finally closing a circle I'd left half-drawn for far too long. I had held Mars in my hands once and now I could hold it again. And if I can convey even the smallest portion of the unfathomable joy I feel about this book, these words, this glorious scrapheap of a planet, then the journey to this point will have been worth it, because Mars is yours as much as it is mine.

PART ONE

EXPLAINING MARS

DAWN ON MARS

It's cold. So cold that the frost glittering on the rocks has been wrung out of the air itself, frozen and turned to ice. The Sun won't rise for another hour, but it's already light enough to see by, although too diffuse to cast shadows. Above you is a sky of pale pink and pastel blue, shot through with white, hair-thin streamers that might be high clouds chasing away from the dawn.

The landscape is flat: not sheet flat, but plains flat, badlands flat. It's rough and rubbly, dominated not by the few variations in height, but by its sheer breadth and width and its utter emptiness. The distance to the horizon is an illusion you've not yet got used to: you think you can see forever, but it's not even the distance of a parkrun. Still, you know, because you've seen the maps and the satellite images, that from where you stand there's nothing between you and the frozen north pole but an unending, monotonous flatness, punctuated by craters of every size that have been gouged out of the ground. The geography only begins to change far to the south, with a planet-girdling slope that climbs inexorably upwards for a hundred kilometres until it plateaus on the pockmarked highlands. Also beyond your closely curved horizon to the south-west, out of sight but looming heavily on your mind, is the land between the red soil and the bright sky where the giant volcanoes squat. They manage to be vastly broader than they are tall, but they still grow so high that they reach above even the clouds.

You feel light because you are light – a third of your usual weight. Walking is difficult. You're used to a falling gait, a motion that rocks you from right to left and back again, heel to toe. Using it here is not just inefficient but unwieldy. You try a few steps, as if you're a toddler again, and find yourself slow and unbalanced. Then you remember what they taught you and move into something halfway between a skip and a lope. You spend far longer in the air than you ever thought possible. Bringing both feet down within seconds of each other, you push off again for another metres-long stride. It's too similar to the dreams you've had of running without touching the ground to be entirely comfortable.

The soil is brittle and it snaps and squeaks as you press your deeply ridged boots into it. The dust that coats the planet from top to bottom is clay-fine, almost oily. Where it becomes loose, it drifts in persistent, dirty puffs and sticks to everything. Your bright white spacesuit is no longer pristine and it'll never be clean again. You wear an ochre cast, deepening in colour towards your feet. It's your badge, your sign, and it'll mark out your stay in days and weeks as it darkens.

The rover you're going to be using this morning is both basic and robust. It's little more than an oversized electric go-cart, designed so that as little can go wrong with it as possible. Its frame is made from open struts of lightweight – but strong – alloy, and its four wheels are independently powered by the fuel cell slung underneath the driver's seat. Each wheel is as tall as you are, and the treads around the circumference consist of hundreds of ridged metal plates. Each plate is sprung to provide suspension, and the amount of springiness is controlled by a computer, allowing the vehicle to navigate soft sand as well as hard rock and ice. The cargo you're driving is strapped to the back of the chassis, already covered in a thin film of dust.

You swing yourself up easily – climbing in one-third Earth's gravity isn't a problem – and settle in the driver's seat. The controls are designed to be foolproof. There's a steering wheel, a trigger to accelerate and brake, and a switch to put the whole vehicle in reverse. The screen in front of you tells you where you are and, critically, how much power you have left. You know that if you abuse that, you're going to be in real trouble. Someone will have to come out and find you before your air runs out. And perhaps they won't make it in time. The only thing keeping you alive at the moment – and every moment you're outside – is the suit you're wearing. Your life is measured in the minutes of air on your back and the watts in your batteries. If the tanks run empty, you will die quickly. If the heater circuits fail, you will die slowly.

You start up the rover and steer steadily north. Compass directions are by convention only, because there's nothing for a magnetic compass needle to find here. All your navigation is done by reference to the small constellation of satellites in orbit overhead. The landscape itself gives very few clues and you can see so little of it anyway. The best way back home is to follow your own tyre tracks, which will persist for a few weeks before being covered by wind-blown dust.

You suppose you're closing on your intended destination. The ground becomes more uneven, with broken blocks of stone buried beneath the subtly ridged terrain. If you go too fast, you could hit one of these sharp-edged rocks and break a tyre-plate. You've been here for less than a week and already the metal of the wheels is starting to look pitted – not just mechanical wear, but chemical erosion from being in contact with the corrosive soil. So you slow down and make a special effort to steer around the most obvious obstacles protruding from the surface – a surface that seems to have flowed in the past, as if it was once thick mud that simply lost momentum and froze in place.

Your position in your seat makes you realise that you've been driving subtly uphill for a while. The terrain is increasingly chaotic, as if something has struggled its way out of the plain and dragged the subsurface with it. There are ridges of rock, hard hills emerging from the pitted plain, fringed by skirts of frost-damaged rock and curtains of landslipped debris. The going is tough. You find yourself driving wheel-first into shallow hollows and losing sight of the horizon until you claw your way back out again.

The gradient gets steeper. There is more rock – not just broken, but shattered, loose and open-jointed, both fragile and lethal. There are cliffs that you cannot possibly ascend, but there is also a way up – not exactly smooth, but passable with care. There are banked ridges to contend with, like driving up concrete steps, but finally, at the limit of your range and the capacity of the rover's fuel-cell battery, you reach your destination, the place to plant your weather station. It's just one part of the expanding network of remote data collection sites designed to feed the appetite of scientists back on Earth. Tomorrow, there'll be another.

You climb down from your seat and look out over the crater that stretches before you. You can't see the other side, and the effect is such that you seem to be standing at the edge of a kilometre-high curving cliff with no end. The ramparts are over-lapping arcs of wall where collapse and time have carved out bites and thrown the digested remains at their feet. You could, you think, plot a course that would get you down there, but not today and not from where you are now. You make sure to stand away from the edge and not make any sudden moves. Even if you sur-vived the fall, you wouldn't live long enough for the climb back up.

The crater floor has more of the same pitted terrain as the landscape you've just driven through, together with sinuous lines and grooves that superficially look like dried-up riverbeds. There are sheltered lees where dust has accumulated into low, rolling dunescapes. In the far distance is a hint of a mountain: triangular, jagged, with more than a hint of archetype about it. This central peak is nearly a kilometre tall, base to summit, standing alone.

The Sun, a cool pale disc, sends streams of light over the edge of the crater wall, progressively illuminating the exposed layers of rock as it rises; the deep shadow and bright light are enough of a contrast to make you squint and wait for your eyes to adjust. Where the sunlight hits, the temperature soars from its night-time low of minus seventy. The frost turns to fog, twisting and swirling, quickly vanishing, and the dust-covered surface glows its traditional rust-red. Behind you, low on the horizon, the bright spark of a moon traverses the sky, its motion peculiar and unsettling.

This is Mars, as it is today, at this hour, this moment. This is Mars, dominated now by ice and time – not inactive, but its processes moving at a glacial speed.

Later in the afternoon, insubstantial winds will pick up and blow the fine dust around half the globe, a distant echo of when those same winds could tear and howl. Where there is now ice, there was once an ocean. Where there is now pink haze, there were once black clouds of volcanic ash. Where this crumbling, moth-eaten crater now sits, the roots of mountains once shook with the violence of impacts so great they threatened to break the planet apart. Mars was built from stories such as these, chapter after savage chapter, creating an entire, glorious world out of grains of dust and starlight.

Mars is unique and everything about it is extraordinary. Even – especially – its future.

MARS AS AN UNRELIABLE NARRATOR

We know what Mars looks like. We can measure it and take increasingly detailed photographs of it from space. We can send probes to the planet and scrape our way into its surface rocks. Our robots can sniff its air and taste its dust. We can find out how hot or cold it is, discover whether it is wet or dry, hard or soft, made of frozen lava or washed sand. We can admire its mountains and its plains and its canyons, vast and broad and deep.

But the moment we ask 'How did it get this way?', we stand on more uncertain ground. There are different routes Mars could have taken to reach the same point – roads that are equally possible – but we don't know which one it actually travelled, or even whether it travelled just one. What we see might be the result of many paths coming together to produce a seamless whole. Each theory, each route to the present configuration, gives us the same Mars.

This presents a problem to anyone who wants to tell a story like this, and to anyone who wants to listen to it. Anything I offer you will look like the truth if I don't tell you about the other possible options that could also get us to where we are now. We can point to an object on the planet's surface – a feature, an anomaly – and while we can know a very great deal about it, we sometimes can't say for certain how it got there.

This is not because of scientific inaccuracies or mismeasurements, but because there are times when several proposed scenarios fit the data well enough to be plausible and none of them can be ruled out.

The history of Mars is drawn not just on its surface but also down into its broken bedrock and up into its frigid air. Most of all, it stretches back into deep time, where the trackways of the past have been obliterated by later events: there's no discernible trace of where they started from or how they travelled, only where they ended up. The history of Mars is simultaneously obvious and hidden. We have to unravel it, see through each layer, to determine the order in which it was made. And sometimes we'll get it wrong.

We're going to find plenty of examples of this. Add to the mix that whatever I tell you could conceivably be true at the moment, to the best of our knowledge, but that by the time you read these words they could have been completely debunked. Science progresses and it doesn't respect reputations. Sometimes I think it gleefully trashes them, leaving a trail of broken hearts and shattered careers in its wake.

So to be plain with you, this book is not going to tell the whole story. That's not to say I won't try, but when I come across alternative explanations of the same phenomenon, I may pick my favourite. Everything here is accurate, but I'm not going to pretend I'm impartial. Mars will always remain over-explained, rather than lacking any explanation at all, and I am not here to stab you with facts and leave you bleeding by the end of chapter three. I'm here to tell you about Mars, with all its ambiguities, and about the times when, honestly, I think it's lying to us.

So rather than me being your unreliable narrator, I want Mars to take that role. Rather than putting the Martian story on rails, I want to open the map out, creases and all. I want to show you Mars, on its own terms, whichever of its stories we decide in the end to believe.

WHY IS MARS SO DIFFERENT?

Physics and chemistry are universal. This is good, because we have a common language to make sense of the results we get from our experiments, whether we do them here on Earth or on Mars.

In a straightforward way, we can talk about Mars's gravity as being just over a third of Earth's – on Mars, you'd weigh a third of your current weight. A Martian day is properly called a sol. In a strange and completely unconnected coincidence, a sol lasts just thirty-seven minutes longer than a day on Earth, but a Mars year is 668.6 sols long, as opposed to our 365.3 days.

But there are times when comparisons with Earth are going to be of limited use because conditions on Mars are so different. Where the differences really begin to bite is when we try to describe what Mars's surface is like, both now and in the past. There is no getting around this, so I'm going to use water as both our example and our guide.

We already know that the temperature at which water freezes or boils depends on air pressure. At sea level, at one standard Earth atmosphere, which is roughly one bar (or more accurately, 1,013 millibars), water freezes at 0°C and boils at 100°C. But we've heard stories from mountaineers that at the top of Everest, where the pressure is only 300 millibars, water boils at 70°C and getting a properly brewed cup of tea is impossible.

We might also have, or know someone with, a pressure cooker. Food cooks faster in one of these because the internal pressure can reach 2,000 millibars – nearly two standard Earth atmospheres – and water will only start to boil when it reaches 121°C. The freezing and boiling points of water are far from fixed.

What's going on is this. The water we interact with every day generally comes in three states: solid, liquid and gas, which we call ice, water and steam. What determines water's state is the strength of the bonds between one water molecule and the next. The bonds between the molecules are strongest in ice, weak in water and non-existent in steam. An increase in temperature means that the bonds vibrate more and eventually tear apart. An increase in pressure keeps the molecules closer together and makes the bonds more resistant to breaking. That's why, at lower pressures, water boils at lower temperatures, and vice versa.

Off Earth, on other planets, moons and asteroids, the conditions for liquid water are rare. At the very highest pressures, water will almost always be in the form of ice, no matter the temperature, which sounds preposterous, but it's true. At the very lowest pressures, including in the vacuum of space, the only possible states are ice and water vapour, with no liquid phase between. Rather than condensing into a liquid and then freezing, the vapour turns directly into a solid, a process known as deposition, and rather than melting and then evaporating, the solid turns directly into a gas, which is called sublimation. The Celsius scale is peculiarly parochial to Earth and it obscures the fact that for most of its occurrence across the universe, water is either a solid or a gas, not a liquid.

The knowledge to hold on to here is that all matter, all physical stuff, responds differently to changes in temperature and pressure. The major component of the Martian atmosphere is carbon dioxide, and that too can freeze solid to form ice if it gets cold enough – much colder than Earth's air – but at most pressures it never becomes a liquid: it only moves between being a solid and being a gas.

Likewise, cold rock is brittle and it shatters when struck or stressed. Very hot rock melts and forms a liquid, which we know as lava when it's above the ground and magma when it's below. But rock under extreme pressure – tens or hundreds of kilometres underground – is neither solid nor liquid. Instead, it becomes plastic: that is, when pushed or pulled, it flows very slowly, like soft wax. It moves only centimetres a year, but if Mars has an abundance of anything, it's time.

The chief differences between our planets, though, are purely physical. Mars is slightly more than half the diameter of Earth and it weighs ten times less. Mars's moons are so small as to be inconsequential to the behaviour of the planet they orbit, while Earth's ridiculously oversized Moon has consequences far beyond that of twice-a-day tides. Mars orbits the Sun at an average distance of 228 million kilometres, while Earth lies closer in, at a distance of 150 million kilometres. These three things – the size of Mars, its moons and its distance from its primary star – have influenced every moment of Mars's history, and will also determine its future.

MAPPING MARS

In order to make sense of Mars, we need an actual map: one that pins features to its surface and gives us fixed points of reference. Unlike the maps found in the front of epic fantasy books, we won't be visiting every last place that has a name, but like an epic fantasy, some of the places we will end up going to have difficult-to-pronounce names. I can only apologise for that.

Making maps of Mars has a long and somewhat ignoble history. Accurate maps – ones that might help us navigate its surface – have only been around for the fifty years since the Mariner 9 space probe took the first detailed pictures from orbit in 1972. Before that, much of what was drawn was not just inaccurate but fanciful and frankly wrong. Credit where it's due for the attempts made by Johann Mädler in 1840, Richard Proctor in 1867 and Giovanni Schiaparelli in 1893: they used serious scientific observations to create their maps, but their technology – ground-based optical telescopes – was simply too limited to pick out anything but Mars's brightest and darkest features.

After Mariner 9 had taken photographs covering most of Mars, the process of making the first proper map could begin. The map-makers divided Mars up into thirty quadrants: eight rectangles above and eight below the equator, six circling the north pole, six around the south pole and the two polar regions themselves. Any attempt to represent a spherical object on a flat sheet of paper is going to distort it, so the polar quadrants aren't rectangles but are instead circular, and each quadrant varies in the size of the area it maps, from nearly 7 million square kilometres at the north and south poles to 4.5 million for the equatorial quadrants.

Each of these quadrants needed a name other than the prosaic nomenclature of MC-1 (short for Mars Chart) to MC-30. These thirty names became wedded to the areas they describe, and they roll off the tongue like a shipping forecast for another planet. In order, from north to south and west to east: Mare Boreum, Diacria, Arcadia, Mare Acidalium, Ismenius Lacus, Casius, Cebrenia, Amazonis, Tharsis, Lunae Palus, Oxia Palus, Arabia, Syrtis Major, Amenthes, Elysium, Memnonia, Phoenicis Lacus, Coprates, Margaritifer Sinus, Sinus Sabaeus, Iapygia, Mare Tyrrhenum, Aeolis, Phaethontis, Thaumasia, Argyre, Noachis, Hellas, Eridania, Mare Australe . . .

The list of names was compiled from the older maps of Mars, and each quadrant was given the name of a principal albedo feature within that quadrant (albedo is the visible light-or-dark contrast). These names subsequently transferred to actual geographical objects on the new maps: craters, volcanoes, plains. As the maps became more and more detailed with each successive mission, they became increasingly populated with new names, many of them informal, and the International Astronomical Union became the final arbiter of nomenclature.

The quadrants have more or less served their purpose: Mars maps are now as detailed as any atlas, but the language of Martian features is idiosyncratic, a reminder of when the learned of every nation spoke Latin to each other. Astronomers in the past used it and now we're stuck with it.

A small flat area is called a palus. A larger one is a lacus. A more extensive low-lying plain is a planitia. A whole region is a regio, and for the want of a superlative, the only use of vastitas is where it describes Mars's extensive northern lowlands. Rugged highland plateaus are plana, or if they are extensive enough, terrae. An individual mountain is a mons, unless it's smaller and more dome-shaped, in which case it becomes a tholus. Neither mons nor tholus refers specifically to volcanoes, but on Mars most of them are. A mountain range is a montes, while hills are colles. A mensa is like a montes, but it has a flat top and is bounded by steep cliffs. Sometimes these cliffs are rupes and sometimes they're scopuli.

Impact craters are still called craters, unless they're paterae, which are craters with scalloped edges as if they've been torn carefully from a sheet of paper. Many of the structures that are paterae aren't impact craters, but collapsed volcanic craters. Fortunately, a volcanic patera is usually found on a mons or a tholus of the same name, so it's normally clear which is which. A chain of craters, however it's formed, is a catena.

A valley is a vallis, but it doesn't have to have been formed by water. A sinus looks like a bay, but it isn't by a lake or a seashore. Insulae are islands, but they aren't surrounded by the sea. A fluctus, however, was almost certainly formed by water – land after a flood. We can also add chaos here, which is heavily broken, unsurprisingly chaotic terrain that may be related not just to a flood but to a dam burst.

Cavi are steep-sided hollows formed by collapsing

underground structures. A chasma is longer and deeper. A fossa is a crack in the crust caused by tension, but they rarely appear alone, so together they are fossae. Where valles, chasmata or fossae cross each other in a complex maze of faults and fractures, they become a labyrinthus. Land that has been squeezed so that it forms linear ridges is called dorsa. Martian wind gathers up dust and sand to form dune fields called undae, and Martian ice reveals itself in the tile-like landscape of tesserae.

Another problem with mapping Mars: where to put the prime meridian, the line of zero longitude? Every map needs a fixed reference point to mark where everything starts and finishes and to act as a ruler for measuring the planet. Competing national interests eventually settled that question on our home planet – the prime meridian passes exactly through Greenwich – but the makers of Mars maps could choose where to place it, and they did so before it was even decided for Earth.

In the 1830s, the German astronomers Beer and Mädler found a small mark just south of the Martian equator, which they denoted 'A'. Schiaparelli continued the tradition with his 1877 map, and the mark was subsequently called Sinus Meridiani or 'Middle Bay'. When Mariner 9 beamed back its photographic treasures in the 1970s, the quadrant-makers needed something more accurate to measure from, so in that same area they chose a small crater within a larger 40-kilometre one. The larger crater was called Airy, after the astronomer who first measured the Greenwich Meridian, and the smaller one, 500 metres across, became Airy-0. After that, with both the ability to land a probe on the surface and the increasing detail from orbital photographs, defining the very centre of Airy-0 became possible. The first lander on Mars, Viking 1, whose location could be pinpointed to the metre, was used to fix precisely where the centre of Airy-0 was, and this became Mars's prime meridian. If Viking 1 was at 47.95137 degrees west, then Airy-0 was exactly at zero degrees.

A further problem map-makers have encountered is how to measure the relief of the land. On Earth the oceans are connected, and this gives a roughly consistent line (which we call sea level) from which to measure the height of the land or the depth of the sea. This line is then refined by accurate measurements of gravity and pressure. But on Mars there are no oceans, so there's no consistent level to refer to, and we cannot measure absolute heights or depths without some kind of reference.

Arbitrarily assigning a zero level – the Mars datum – was necessary, but it also needed to be both useful and reproducible. Initially, it was chosen to be the level at which Mars's air pressure matched the triple point of water: the unique lowest pressure where water can exist as either gas, liquid or solid. This is just over 6 millibars and it was entirely reasonable for Mars, but later on, more and better data led to the Martian gravity map being used to define the datum. The variation in the path of a satellite orbiting Mars can be used to calculate the planet's centre of mass, and from there we can measure the exact average radius at the equator. Everything that is above this imaginary radius is above datum; everything beneath it is below datum. The highs and lows of Mars can now be measured. Its highest point is the summit of Olympus Mons, 21 kilometres up, and its lowest is the bottom of the Hellas crater, 8,200 metres down.

Armed with this bundle of information, we're ready to write the first chapter of Mars's glorious history.

PART TWO

BEFORE THE BEGINNING

(More than 4.5 billion years
before the present)

THE GIANT MOLECULAR CLOUD

If I'm going to explain Mars to you, I have to go back to the time-without-time, before the solar system existed. Reaching back beyond 4.5 billion years is going to be stretch, but what happened then is still apparent now. It's all connected, a long chain of unbroken links.

Everything that Mars was and is started as a cloud of cold smoke, just like everything else in our solar system: our Sun, the planets that orbit around it, the accompanying panoply of smaller objects and the atoms that make up our own bodies. It was a cloud so vast that it stretched for 600 trillion kilometres and contained enough matter to form thousands of individual stars.

This cloud was made from the sweepings of the galaxy itself, the leftovers from its own formation and the detritus of earlier giant stars that had already burned themselves out. In the space between the matter-packed spiral arms of the galaxy, gravity was a negligible force, but it was not nothing, and so a tenuous drift of stuff coalesced. The cloud could just about keep itself together by merit of not being close to any object large enough to disrupt its fragile existence, but it still had a limited lifespan: it might fade away without consequences, it might collapse and give birth to a few massive new stars which would then burn off the rest of the cloud, or it could simply wait for the next star-rich galactic arm to pass through and spread it out again.

Despite being thin and filamentary, the cloud was still dense enough to block out starlight. Most of the molecules of gas within it were hydrogen, the default element that matter organises itself into in the universe, but other elements and molecules were present, so sparsely spread that they rarely collided, and so cold that there was no hope of a chemical reaction when they did. Inside, the cloud was almost as sterile and inert as it was outside.

Out in this wide interstellar gap, nothing existed that could stir the cloud. But for this story to be more than just a prologue, we need an initiating event, something to start the clock and begin the count. The shockwave from a supernova is one of those things.

The larger a star is, the faster it consumes its nuclear fuel, and the largest of all will attempt to burn their own ash – helium, carbon, neon, oxygen, silicon – until their cores collapse abruptly under their own weight, like a tower falling even before the upper storeys have registered that anything has happened. This triggers a supernova, the sudden brightening of a dying star. Spectacular from very far away, they are terrifying and deadly close up – close being a relative term; life on Earth would be severely impacted, if not ended, if one occurred within thirty light years of us.

For our interstellar cloud, the supernova shockwave came in two parts. The first was a blinding flash of intense energy, enough to bathe surrounding star systems in lethal radiation; the second, lagging behind, was a wave of matter, moving away from the stellar corpse at speeds of up to 10 per cent of the speed of light. A wall of particles – including exotic heavy elements that are rare in a universe dominated by hydrogen – slammed into our molecular cloud and mixed with it. Although launched from billions of kilometres away, the impact was enough to create local variations in density within the cold, inert smoke.

For the first time in millions of years, the gas began to move. Gravity's weak grasp started to strengthen and the cloud broke up into discrete, albeit still huge, trillion-kilometre clumps. As these clumps spread out from their original positions, the subtle movements of the gas warped them into shrinking, rotating discs of material, each capable of birthing at least one star, and potentially a planetary system of its own.

Inside our unexceptional disc, matter falling from the edge towards the centre delivered energy in the form of heat as it collided with other particles. Matter interacted with matter. It stuck together. It began to pool, and over the next hundred thousand years, the initially thin, almost abstract gas formed a structure at the centre, dense and warm. That increasingly dense middle, where the vast majority of the material ended up, started to run away with itself. The more gas it attracted, the hotter and denser it became. Rather than the heat stealing away into space, energy was trapped behind the layers of falling, compressing dust. The central mass turned thick and dark: this is what would become our Sun.

Outside, the leftovers formed a thin rotating disc of matter: the protoplanetary disc. From this, everything else was fashioned: the whole halo of objects that circle our Sun, from the closest to the furthest away. The flattened disc stretched for 30 billion kilometres from the centre. Collisions of material within the disc were common, and as the disc heated up, the lighter elements – hydrogen and helium – were pushed to the outside, while the heavier elements became dominant in the mid-plane of the disc and towards the centre.

Chemistry happened. Where present, solid metal and silicon oxides condensed directly out of the gas, using oxygen to form stable molecules. Heat fused particles together. There was dust now, grains of matter, and these grains coalesced and melted and grew larger to become chondrules – millimetre-sized blobs of molten silicate glass. Chondrules were the basic, if tiny, building blocks of the solar system, forming and reforming during collisions over the next three million years, trapping material that might otherwise have been pushed to the edge of the disc. We find chondrules in meteorites today, frozen into the rock, packed together with dust and fragments from the protoplanetary disc.

As further shockwaves rippled through the material, these chondrules stuck to each other and collected dust on their molten surfaces. They collided. They clumped. As they grew larger, they experienced more drag against the gas they encountered: as their speed dropped, they orbited ever closer to the centre. Further out in the rotating disc, where it was colder, it began to snow. Fluffy ice crystals made of volatile, low-boiling-point compounds stuck to each other, creating mobile drifts that slowly orbited in the gas-rich extremities.

It's at this point that things become difficult to fathom. If Mars is an unreliable narrator, the solar system speaks of itself in fragmentary, cryptic phrases. We know what we ended up with: a yellow star, a quartet of rocky inner planets, a gap filled with scant rubble, and then two giant gas planets, two giant ice planets and a corona of ice and dust. But the processes involved in getting there from our collapsing dusty disc, thick with hot glass beads and tiny, fluffy snowballs? It's a question without a definite answer. There are signposts we can follow, though, and some of paths they lead us down are quite extraordinary.

ACCRETION

Accretion is the gradual accumulation of solid material into ever larger masses by a continual and cumulative process of low-speed collisions. It's what we assume happened in the protoplanetary disc, but there's a series of unresolved problems with the process that happen at key stages and ought to have prevented accretion from happening at all. And yet, we have planets. Clearly something is wrong – or right – and we simply don't understand enough about what happened.

The first problem is this: we know that if we push ice together, especially when it's in the form of fine, crystalline snow, it sticks. A gentle amount of pressure causes a slight melting of the ice, and the two clumps of snow become one. The same can't be said of glass beads. Chondrules and collections of chondrules, no matter how hard you mash them together, don't form fist-sized rocks. Quite the opposite: two pebbles thrown together are likely to break apart into several smaller fragments, not create a larger one. This is called, without irony, the bouncing barrier.

Once an object within the protoplanetary disc grew large enough, roughly a metre across, it created its own gravity gradient and so could accumulate extra material to grow larger still. But getting to this boulder stage from mere pebbles was fantastically difficult. We think this happened gradually, piece by piece, but we don't know how for certain. Perhaps the region where

these metre-sized rocks formed was hot enough for the chondrules to become sticky. Perhaps snowballs falling in from the
outer disc managed to gather sufficient rocky material to form
boulders before the ice evaporated. Or perhaps it didn't happen this way at all: maybe accumulations of chondrules formed
clumps and travelled around inside the disc as nothing more
cohesive than a loose train of gravel caught in an eddy of gassy
dust, which slowly accumulated enough mass to begin accretion
through gravity.

Having swerved past one problem, we move seamlessly
into the arms of another – something called the drift barrier.
Once these small snowy or rocky boulders formed, they became
large enough to experience aerodynamic drag from the slower-
rotating gas within the disc. Thus slowed, they spiralled inwards.
If they'd ever reached the central mass, they would have fallen
into it and been lost forever to the surrounding disc. Indeed, in
the complex computer models used to simulate the protoplanetary disc, all metre-sized boulders that form in the flattened
extremities do get quickly sucked into the dense dark region at
the centre. If this had happened, the disc would have become
sparse and depopulated, with nothing left to build planets from,
so both the bounce barrier and the drift barrier must have been
overcome somehow.

As more material accumulated in a decreasing number of
larger objects, the dust and the gas in the disc itself thinned out.
Collisions that might have shattered a smaller mass now grew
them as gravity played its part. Larger bodies, now kilometres
across – planetesimals – swept up the remaining debris and each
other.

Beyond the snowline – the radial distance from the centre of the disc beyond which frozen water, and other gases cold enough to, condense – planetesimals were composed of loose accumulations of chondrule-rich rocks and ices and, trapped within the mass, actual gases. These are the classic dirty snowballs we now call comets, and they dominated the outer part of the disc. Closer to the hot centre, ices were unstable and the planetesimals consisted of metal-rich silicate rocks composed entirely of chondritic material. This gave us the first major division in our solar system: rock and metal in the inner solar system; rock, ice and gas in the outer.

In a mere three million years, the disc had evolved from slowly rotating smoke to thousands of barrelling mountains of rock and ice. At the centre, where most of the matter was, the journey towards stellar nuclear ignition was continuing as more material fell inwards, heating and crushing what was already there.

So on to a third problem. We had kilometre-sized planetesimals in chaotic orbits about the centre, scouring debris from what was left of the disc as well as devouring each other in increasingly violent collisions. Most of those collisions, large and small, reduced the orbital speed of the planetesimals and therefore their distance to the centre. As a consequence, all our models suggest that it's the innermost regions of the disc that were the densest, and so it's here that the largest planets ought to have formed.

This is what we see when we examine other planetary systems around different stars. When we knew of just one solar system – our own – we thought of it as typical and we assumed we'd find similar arrangements of planets elsewhere. But since our discovery of exoplanets (planets orbiting other stars, and we find them almost everywhere we look), it seems that our particular configuration of planets is far from usual. The average planetary system has planets that are all roughly the same size and evenly spaced apart. The larger the planets, the further they are from their neighbours. And as a rule, other planetary systems form far closer to their star – all of the planets lie inside the orbit of our innermost planet, Mercury.

Clearly something extraordinary must have happened early on in the formation of our own solar system that meant we have the arrangement we see today. The four rocky inner planets are unequal in size: Mercury is half the mass of Mars, which is in turn one-tenth the mass of Earth and Venus. Then there's a gap: the asteroid belt isn't a failed planet – the total accumulation of asteroids in this region weighs less than 1 per cent of Mars's mass. Jupiter and Saturn are distant giants, freakishly huge gassy bodies; further out are Uranus and Neptune, smaller and icier than the other giants but still vast.

Jupiter – along with the other giant gas and ice planets – is in entirely the wrong place. It should be much closer to the Sun, and therefore the rocky inner planets shouldn't exist at all. Mars should, by rights, be the same size as its two closest inward companions. There should, perhaps, be a fifth rocky planet of similar mass beyond it, where the asteroid belt currently is. But this is not the solar system we have.

The odds of our solar system turning out the way it did are around one in a thousand, and potentially even smaller than that. Any working theory of the formation of solar systems has to account for so much strangeness. We reach for answers and they elude us.

PLANETARY EMBRYOS

Three million years after the collapse of the giant molecular cloud, the solar system was ready for its final phase of planetary growth. The protoplanetary disc had evolved from being dominated by gas into one dominated by discrete lumps of matter – a debris disc.

At its centre was a swollen bolus of heat that accounted for almost all the gas from the earlier cloud. This was the protostar – it had all the mass of the Sun, but the nuclear fusion that would make it shine for the rest of its life had not yet begun. As the shrouding dust cleared sufficiently to let light pass through, the protostar was finally revealed in its monstrous glory: balefully huge and blindingly luminous – four times larger and sixty times brighter than the Sun is now.

As the protoplanetary disc transitioned into its debris disc phase, the infall of supersonic gas, dust and high-boiling-point compounds reduced, and the outside of the protostar cooled. From being in balance – where the urge to pull its surface inwards due to gravity was equalled exactly by the pressure of the heat pushing everything apart – the seething, roiling ball started to shrink. The matter at its very core was crushed ever closer together: lithium atoms were torn apart and hydrogen formed from helium, liberating enough energy to slow the contraction, but not stop it. The process took millions more years and would end with the full-scale nuclear fusion of hydrogen.

Orbiting this hyper-luminous protostar were the tens of thousands of kilometre-sized planetesimals. It was a chaotic time, and collisions were frequent, relative speeds low and accretion rates high. Planetesimals attracted each other even if they avoided collision during any particular pass: their gravities interacted, pulling each other into new orbits that could bring them into contact the next time around. Some experienced runaway growth, absorbing all the other planetesimals they encountered. This quickly led to fewer, larger bodies that collided only rarely, and when they did, it was with greater energy. Fragmentation, melting and mixing became important for the first time.

But where in the debris disc did this great accumulation happen? We know from our models where it ought to have been – where the greatest density of matter was, close to the primary star – and it's reasonable to assume the same for our own solar system. Our early rock-based planetesimals all orbited in a narrow band close to the centre, a doughnut shape with well-defined borders. The inside border was close to the protostar, far inside the present-day orbit of Mercury: any closer and it would have been so hot that even metals and rock would have boiled off into gas. The outer border for rocky planetesimals was no further than where Earth currently orbits; beyond that there were insufficient bodies to accrete with.

Icy planetesimals stretched that border outwards somewhat, but as they collided and grew, they were subject to the same forces as their rocky counterparts and they moved closer in. The whole of our solar system's planetary formation process became squeezed into a comparatively busy and narrow region. Somewhere in there was Mars – or rather what would become Mars, a proto-Mars that existed as a thousand-kilometre-wide planetary embryo at this point, along with the other few hundred planetary embryos formed by runaway accretion. They began to dominate their own particular orbits. They collected other stray planetesimals and increased their size without being deflected from their paths.

We didn't end up with a few hundred planets. While these embryos formed the apex in the hierarchy of the rapidly clearing debris disc, there was a little way to go yet. Proto-Mars, along with proto-Earth, proto-Venus and proto-Mercury, was already present, as were dozens of other embryos that never managed to gestate to term. Critically, just on the other side of the snowline from proto-Mars was the monster that killed most of them: Jupiter.

Together, Jupiter and Saturn represent 90 per cent of the planetary mass of the solar system. Jupiter alone accounts for 70 per cent. It's about as big as a planet can get – if it had more mass, gravity would make it denser and therefore smaller in size. Forming where it did, near the snowline, it started as just one of the planetary embryos in the debris disc. Roll the dice a thousand times, and in nine hundred and ninety-nine cases Jupiter would have become a reasonable-sized rock-cored planet, covered in layers of ice. But we live in that exceptional timeline where Jupiter voraciously consumed mass – as well as attracting ice and rock, it was still ploughing through the remnants of the dust and gas that were more abundant in the outer edges of the disc. It grew into a body that was over a hundred Mars-equivalents.

As it grew, Jupiter slowed and spiralled inwards. As it crossed the current orbit of the asteroid belt, it gobbled up everything in its path: planetesimals and planetary embryos alike, material that should have coalesced with proto-Mars but now never could. Earth and Venus were still relatively well supplied, while Mercury – half the mass of Mars – appears not to have grown much beyond the planetary embryo stage. Objects that Jupiter failed to absorb, it flung around the disc. Saturn was briefly left behind at this point, but in the absence of the ever-inward-moving Jupiter, it too started to accumulate gas. It grew rapidly and followed Jupiter towards the inner region of the disc.

There was potentially nothing to stop Jupiter's inward march. We see planets of similar size around other stars, parked in very close orbits, where the heat of the star strips away their vast reserves of gas. These 'hot Jupiters' have a limited lifespan. They eventually become naked cores, huge baked cinders of planets perpetually roasted in the intense starlight. But Jupiter did stop. More than that: it moved away again, and it dragged the rocky planets back out with it.

Orbital resonances are odd things. At first glance, two planets orbiting a star seem so distant from each other as to be unable to affect one another's motion at all. And yet, planets and moons can and do fall naturally into cycles, the subtle drag of gravity pulling bodies towards each other as they pass by. For much of their formation period, Jupiter and Saturn had a 3:2 resonance: for every three orbits of Jupiter, Saturn made two. When Jupiter moved inwards, that resonance was broken.

But when Saturn fell sunwards too, it reattached itself to the 3:2 resonance frequency. Jupiter, then located around the current orbit of Mars, started to move out again, caught up in its resonance with its smaller outer companion. Jupiter retreated to beyond the asteroid belt once again, scattering debris across the solar system for a second time. Saturn's presence pushed Uranus and Neptune further out, into the icy debris at the very edge of the disc, and possibly caused a hail of distant snowy planetesimals to arc inwards towards the more-populated centre.

Giant planets swinging backwards and forwards through the early solar system like sailing ships, knocking rocks aside as they went, sounds preposterous. And yet the fine mill of computer simulations shows that this scenario is not just possible but probable. We have to get here from there, and it seems there are very few ways to do it. Can we explain the low mass of Mars? Why do we have an asteroid belt? Where are the missing terrestrial planets? Why is everything so far away from the Sun? How did Jupiter become so vast?

This theory – the Grand Tack theory – is our current best guess. If our calculations are correct, it was a startlingly swift process: Jupiter's innermost incursion, reaching the current orbit of Mars, took place a mere 100,000 years after it formed, and it moved back out to its current orbit in just another 400,000 years. There are variations on the Grand Tack theory, and there are other theories that end up in the same place (of course they do, otherwise they'd be rejected), but in the Grand Tack's favour, it continues to give gifts long after it's finished.

For now, we have marshalled everything into position. Our solar system is complete and all the planets are in their orbits. Proto-Mars is at the right distance from the protostar; it has the right mass and the right chemical composition. Mars the planet is ready to be born.

PART THREE

EARLY MARS

(4.5–4.1 billion years before the present)

MARS AT THE START

You can't stay long here, but there is a kind of awful majesty to a landscape that shimmers with so much heat; it's like staring into the heart of a furnace. Spacesuits are useless: not only is the ground hot enough to melt anything resembling a boot, the air pressure could crush you like an empty can of pop. You're in a heavily insulated bathysphere-type structure; the walls are half a metre thick but still you can hear the metal around you tick as it tries to simultaneously expand with the temperature and contract under the external pressure. The one saving grace is knowing that if the hull fails, it will fail faster than your nerve endings can carry pain.

The tiny porthole you have to make observations through reveals this: a flat, black plain. The surface is sometimes ropey, as if a liquid has congealed at the same time as flowing, and sometimes blocky, as if a solid crust has been tipped and broken and tipped again. From orbit this seemed to be the safest place to land, but now you're here you're not sure how stable the rock is beneath you. There are other places riven by deep rents through which the heat from Mars's formation still glows. This broad, dark, cinderous plate, though, has remained uncracked long enough for you to risk a momentary touchdown.

A little way distant is what you've come to see – and measure – and you can feel it through the landing pads, despite the shock absorbers and all the layers of ceramic foam between you and the surface. The vents that roar out gases into the Martian atmosphere from deep underground – mostly superheated steam, boiling carbon dioxide, oxygen and hydrogen – are so hot and fierce that the already oven-like air twists and braids. You know they're there, but you can't actually focus on them. Each vent is a mirage, albeit one that would sear your flesh and then rip it off with the force of its discharge. The newly formed land is scattered with these sites: cracks and tubes that channel pockets of gas from under the kilometre-thick layer of cooling rock and squirt it out into the dense, bright air.

The horizon is a line made inconstant by the rising heat. The sky itself is a luminous white, cathedral-high, with a complete cloud cover that roils like the thickest smoke. The top of the atmosphere is at a ridiculous height, almost 10,000 kilometres up, and it's only close to halfway up that the water vapour becomes cool enough to condense out. Somewhere up there it's raining, but that rain will evaporate long before it reaches the ground.

You check the temperature outside, and you have to look twice because you think the sensor might be broken. It's registering 350°C and a pressure of 200 bars. The probes you want to launch have already been pressurised to this. You don't expect them to last longer than a few minutes, but you hope it'll be long enough to take at least some readings.

You've been in the vacuum of space. You've been at the bottom of the deepest oceans. You've been on the flanks of an erupting volcano. But you've never before been in a place so implacably hostile in so many different ways. You drop the probes. You want this done and over so you can leave as soon as possible. If the extreme conditions were all you had to deal with, then you wouldn't be so nervous, but there's always the possibility of a meteorite strike. The early solar system is still thick with potential impactors: you think you've tagged them all and calculated their orbits, but if you're wrong and a big one hits almost anywhere in the same hemisphere as you, it's over.

Three probes scuttle across the baking surface, high-stepping on their spindly legs to limit contact with the ground. They look almost comical, like they're dancing, but this, you're assured, is the only way they're going to make it to their target. They skitter and slide, but their robot brains instantly make all the calculations required to keep them on track and upright.

They're closing in on the vents. All you can see of the probes now is their flickering silver outlines against the dark rock. Then the first one missteps: it trips at the edge of the vent on a piece of forcibly ejected rock, falls into the billowing gas and it's gone, high into the air until it's lost from sight. It's somehow still transmitting, though. You quickly take a sample, even though the probe is only partially surrounded by the gas expelled from the vent. At some point, it will come down again and will probably end up dashed on the rocks below.

You move on to the remaining probes. They extend their proboscises into the almost supersonic jet of gas and sniff. The initial readings come through and you get to see what Mars is made of. The results are useful, but they're not making up for the peril you feel.

Another probe fails: having stood still for sufficiently long to measure the jet, it has now overheated. It's flatlined, fried. The third is still hanging on, drinking deep at the vent when your console pings. Something's coming and you have to go, now.

Take-off is hard. The rockets kick you into your reclining couch and multiples of your own weight hold you there. Outside, through the porthole, you glimpse a distant trail of smoke arcing swiftly downwards, even as your own takes you up. It's difficult to tell how big that particular meteorite is, or how hard it will strike – the thick atmosphere robs all but the very largest of their energy. That it's made it this close to the ground indicates that it's going to hit.

Sure enough, there's a sudden flash that prints an oval of light against the wall inside your cabin. Then comes the snap as the shockwave heaves through the soup-like air. But you've made it: you're high enough and far enough away that the overpressure doesn't crack your hull like an egg, even though you feel the course correction kick in as the strong winds shake your ship.

Down on the surface is a bright circular scar, a glowing lake of molten rock contained within newly formed high walls, and a blanket of rubble that is still spreading across the dark rock of Mars in a ballistic arc. Where the rubble falls, it'll tumble and rattle to a stop, subtly changing sightlines as well as burying the fractured surface under a layer of newly broken, altered rock. The light's already fading as the fireball dissipates over the nascent crater, and then it's gone: you're in the cloud layer, heading back into orbit. You're relieved to have made it out alive and you won't be hurrying back.

PLANETARY MELTING

Barely ten million years after the first shockwave rattled the giant molecular cloud into life, Mars had accumulated all but a final few collisions' worth of its eventual mass. By the measure of deep time, this is only slightly longer than an eye-blink and a little less than a drawn breath.

Mars never experienced what the planets sunwards of it would: repeated, late, massive impacts with other planetary embryos that tore both impactor and target apart, melted and mixed their contents and then coalesced them again into a new, larger whole. Neither did it experience, as is suspected of Mercury, an impact so vast that its outer layers were almost completely stripped away and dumped into the Sun. Mars, stuck at the edge of the rocky-planetesimal-rich band, avoided both the injury and the benefit of collisions with the very largest non-planet objects. Smaller planetesimals still fell on Mars and left their marks, but no great pile of matter was added to it after formation. It might have been left underweight thanks to Jupiter's pocketing of much of its potential mass, but it was also complete.

As it accumulated through chondrule to boulder to planetesimal to embryo, the stuff of Mars was more or less the same throughout. Most of its material came from the surrounding orbits within the debris disc, and it all had a similar chemical composition. Lying just inside the snowline, Mars found itself poor in volatile gases and ices, and rich in silicates and metals; a cross-section through it would have shown it to be a jumble of rock from surface to core, held in a more or less spherical shape because of gravity.

But that would change, and the very start of this story gives us the reason why. The exploding star that shook the molecular cloud and caused it to collapse also pressed exotic matter into it. Forged from the ash of successive nuclear burnings in other, already-dead stars, elements heavier than iron were mixed with the mainly hydrogen–helium cloud. A few of these elements were radioactive, and as these elements spontaneously disintegrated, they gave out energy in the form of fast-moving fragments of atomic nuclei and high-energy gamma rays: radiation. Half-lives are the measure of how quickly a radioactive element splits apart – and an indication of its energy output: one half-life means half the element has gone. Some, like uranium, have half-lives measured in billions of years (and can be used in the radiometric dating of rocks). But others we'd never normally encounter had very short half-lives: a few tens of thousands of years.

A lump of radioactive metal is warm to the touch. That heat is the result of its decay from one unstable element to the next, and the next, until it reaches a stable atomic state. Every little block of material is generating a tiny amount of heat, sufficient to raise the temperature of the whole before it's radiated away. But if we trapped that heat within tens or hundreds of kilometres of dense, insulating rock – which was at the same time also generating its own heat – that radioactive heat would accumulate faster than it could escape.

There is a minimum size for a rocky planetesimal to trap enough radioactive heat for it to partially melt. That size is somewhere between 20 and 30 kilometres across. At this point, Mars was 3,400 kilometres in diameter. Such was the speed of the accretion process, Mars was completed before the heat started to build. But once it did, the temperature rose so high, so fast, that the rock began to soften and sag. Pockets of molten rock formed. Any metal – radioactive metal included – sank to the bottom of each pocket, and because it was falling, it released more energy, which was turned into yet more heat. The whole process took just a few thousand years – a mere moment. Mars melted, spectacularly and completely, from centre to surface.

When we think of molten rock, we have a particular image in our minds – that of glowing, red-hot lava with dark flecks of crust floating on top, running in rivers down the sides of a volcano. But on a planetary scale it has a different quality. On early Mars, everything at the surface melted. From pole to pole, there was a single global ocean of seething molten rock. Geysers of high-pressure gas roared through this ocean and sent fountains of fire into the sky; any volatile ices that had been caught up during the making of Mars boiled out to form the planet's first dense atmosphere of water and carbon dioxide, which acted as a greenhouse-like blanket, protecting it from the vacuum of space. And keeping the heat in the rock and making it stay liquid for longer allowed the magma more time to separate out into temperature-dependent layers.

Below the kilometres-deep liquid surface, the pressure was so great that rock hot enough to melt instead flowed like warm putty. Far denser than the surrounding rock, metals like iron and nickel were dragged downwards by gravity until they formed a vast accumulation of liquid metal at the centre of the planet. The huge amount of energy released by this sinking ensured that the warm putty-rock started its own movement: thousands of kilometres of hot, soft rock rose in planet-wide curls, reaching for the surface, cooling, then sinking again under its own weight. The less dense silicates separated out and floated to the very top, where they melted and formed continent-thick liquid rafts that drifted on denser but still incandescent rock.

What we might imagine would happen next – that the magma ocean froze from the top down, and over long years gradually thickened as heat escaped into the atmosphere – didn't. Certainly, minerals with high melting points started to freeze out as the magma cooled, and the coolest part was at the very top, where the heat escaped into the atmosphere and was carried away. But minerals with lower melting points stayed liquid, and the magma ocean became a kind of hot rock slurry that contained solids but still flowed. And while solids are generally denser than liquids and will sink through them, some liquids are denser than some solids, which can lead to an unstable heavy, cooler solid layer balanced on top of a lighter, hotter semi-molten slurry of crystals.

On other bodies in the solar system, the low-melting-point fraction froze in place as expected and became the crust, but on Mars the conditions were different and we don't know why. There, the deeper layers of rock stayed fluid for longer, and the layer on top – dense and rich in iron and radioactive elements – sank again, potentially all the way back down towards the core. So Mars recycled its first, 'primary' crust, and with it, its payload of heat-creating elements. The result was far-reaching: Mars was going to be able to generate internal energy for far longer and at a higher rate than it would otherwise have been able to do.

Mars turned from an undifferentiated rocky lump, with the same characteristics at its centre as at its surface, into a stratified planet. At the centre was an iron–nickel–sulphur core, rich in radioactive elements, solid in the middle, liquid on the outside; on the outside, a thin secondary crust of frozen rock, laden with iron oxide; and in between, a vast hot, viscous, mobile mantle of silicate rock.

This abrupt, cataclysmic, planet-wide and planet-deep event essentially reset the clock of Mars. Everything that it was before – the records held in its layers of compacted, chondrule-rich rubble and impact-melted rock – was utterly obliterated. This was its year zero. From this time onwards, after the global melting and formation of Mars's new crust, any collision with material still being swept up from the debris disc stood a chance of being recorded and preserved as an impact crater.

CRATER FORMATION

Meteorites come in every size. Even the very smallest grain of sand can pit a surface it strikes if it's moving fast enough – that is, in the order of kilometres per second. The particle squashes against the target rock, compressing both. The moment the moving particle stops, all that energy comes bounding back, focused inwards. The particle vaporises and forms a tiny expanding sphere of gas and molten rock. The mark it leaves is the familiar circular indentation with a slightly raised lip that we call a crater.

Scaling up from millimetre- to metre-sized meteorites, all the features of simple craters remain present – they just get more impressive. Parts of the target are vaporised along with the impactor, forming an instantaneous source of molten rock called an impact melt. Most of this will be thrown out of the crater, together with all the pulverised and excavated material, as a chaotic mix of broken rock, cooling melt and fine dust. The debris splashes out radially around the crater as an ejecta blanket, thickest nearest the crater rim and thinning out towards the margins; the size and depth of the ejecta blanket depends on the gravity of the planet and the material it's made from. The crater floor itself is cracked and crushed, and after the event the crater walls tend to slump inwards, blurring their initially sharp edges.

Kilometre-sized meteorites produce huge, broad craters. The phenomenal pressures in the ground beneath the crater cause the rock below to fracture, at the surface and deep down, and immediately after the initial violence of the compression and rebound that vaporises the meteorite, a lake of molten rock collects inside the crater, filling it and levelling it out. A central mountainous peak often forms, caused by the sudden release of energy. Beyond the crater wall, the surrounding rocks are fractured in concentric circles, which can subsequently slip downwards, slumping into the crater floor and exaggerating the initial diameter of the crater. The ejecta blanket is projected outwards at supersonic speeds and travels as a red-hot wall of debris. The highest-thrown material arcs down with such force that it causes secondary craters.

When meteorites reach tens of kilometres across, the effects of a single impact on a Mars-sized planet are global. The impactor punches down into the crust like the fist of a furious god, breaking the rock, bending the plastic mantle beneath and sending a plume of material outwards across half the planet and potentially beyond: into orbit, and even out of it. The impact forms a basin a hundred kilometres across or more, filled with glowing impact melt that abuts the mountainous internal rings caused by reflected shockwaves. Ejecta buries everything for a thousand kilometres around under a suffocating layer of churned-up debris. Dust blots out the Sun, and secondary craters can occur anywhere over the entire planet as debris re-enters the atmosphere. Even some of the atmosphere itself can be lost, thrown out into space, never to return.

The extraordinary effect and suddenness of these large meteorite strikes is sobering. They don't just have the power to excavate huge quantities of material to great depth and launch it outwards – they can sculpt entire regions, alter the climate and potentially knock the planet off its axis, all within moments.

However, meteorite impacts also supply us with one of our most useful tools for telling Mars's story. If we know, even roughly, the rate at which craters are formed, we can not only date the various parts of Mars's surface relative to one another, but crucially we can give a rough absolute date in terms of a fixed time for a crater's formation, without ever having to set foot there.

CRATER COUNTING

Back in the eighteenth century, when geology was becoming a proper, systematic science, one of the early breakthroughs was the insight provided by the law of superposition. Simply put, this is the idea that younger rocks lie on top of older rocks. Following that observation, we discovered that we could order our own geology in terms of youngest, oldest and the layers in between, simply by comparing outcrops of rocks in various places – as well as natural cliffs and riverbanks, we could use information taken from railway cuttings, canal bottoms and mines.

After mapping all these out, we were able to tell with reasonable certainty which layers were the youngest, and then go all the way down to the oldest in an iterative process, until all the rocks were accounted for.

The Martian landscape might be different to ours, and our access to it very limited, but we can use the increasingly detailed photographs we've taken to map out the various landforms and rock types that are visible, and supplement those images with our knowledge of how these shapes were formed. We can do something similar for Mars as the first geologists did for Earth.

If lava flows into a valley, it has to be younger than that valley – we have no real way of telling how much younger, but if further on the valley appears to cut through a crater, we can tell that the crater is the oldest of these three features. More investigation might show that the crater wall has slumped down later, partially blocking the valley. This event, then, is younger than the valley, but it could be contemporaneous with the lava, or younger still than that.

There are further tells the landscape gives us, and they are all interconnected and overlapping. Two separate lava flows from two separate volcanoes could reasonably appear to be the same age, but by noting the features each of them cross, or are crossed by, we can determine which is older, even though they themselves never meet.

We can tentatively identify other features as being formed by water, and by wind, and by ice, and know how these riverbeds, sand dunes and glaciers are connected. Add these to the structural elements – the areas where the land has been thrust up or brought down, the valleys and mountains, and underlying it all, the ancient Martian crust – and a careful, painstaking examination of the surface geology of Mars reveals a chronology of events.

What would be best of all would be to put exact dates on that relative history. But in the absence of radiometric dating of pristine rock samples, we are left with the less satisfactory but nevertheless powerful tool of crater counting.

Crater counting is a useless exercise on Earth because the planet renews its surface so often: craters here are transient features, eroded not just by the effects of our highly active weather but by the constant motion of the underlying crust. Earth has a system of plate tectonics: its mantle moves in a complex series of gyres, with hot rock rising from the depths and pushing against the crust as it cools, before sinking again, forming vast, ponderous convection cells. The crust in turn travels along the surface, driven by the mantle currents, splitting and moving and colliding, riding up and being dragged down. Mountains rise, basins sink, oceans form and close. Any record of impacts is reduced to the most recent or the very largest, and even then most visible signs become scoured from view.

The situation is different on Mars. While, like Earth, Mars also had a mantle that moved, a full system of plate tectonics was only ever a brief episode – a couple of hundred million years, if that – after which any movement of different parts of Mars's crust ground to a halt. From an active sheath of rock that split and moved and sank, it became a single, stagnant, solid shell. As a consequence, the whole of Mars has a basement of crust that dates all the way back to when it froze out of the global melting event, 4.5 billion years ago. It undergirds every subsequent structure, and the record of every single impact made on it is potentially preserved from those earliest times.

So how does crater counting work? Imagine a big square of wet cement – this is our early Martian crust – and next to it, a bag of unsorted rocks ranging in size from big half-bricks through lumps of masonry all the way to bits of pea gravel: these are our meteorites. We have a shovel, which, at the risk of stretching the analogy too far, represents the number of meteorites that can impact Mars in a given time. We dig that shovel into our bag of rocks, and whatever is on it we fling at the cement.

What we've made is lots of craters. Small craters, medium craters, large craters, in a random pattern over the whole area of once-smooth cement. Let's take another shovel-load of rock and throw that on too, just for good measure. Some of the craters will overlap; some of the new larger ones will have erased some of the older smaller ones. Large craters from our previous effort will now be pockmarked by gravel-sized ones.

Now, we get some more cement – let's call it a huge lava flow – and we cover half the original area. We can tell which is the original surface, because it's covered in craters, and which is the new surface we've added on top, because it's completely unmarked. Now we throw another shovelful of rock high into the air. Some of it comes down on the original surface, scarring it further. Some of it comes down on the new surface, marking it for the first time. But the important thing is that we can still tell the difference between the two – even though the newer surface now has some craters too – by comparing the size and number of craters over both areas.

Let's do it again. Take some more cement – another, more recent lava flow – and cover half the original layer and half the second layer. We now have three ages of surface, but if we throw more rocks at them, we should still be able to work out which is youngest and which is oldest.

This is how we tell the relative ages of the different surfaces: the more heavily cratered a piece of ground is, the longer it's been exposed to cratering events. There are caveats here, though. A surface may become so saturated with craters that any new crater might erase the memory of an older one. This means that two very old layers might record the same crater density, even though they're not the same age. At the other end of the scale, an area might be too young to have any craters at all, but there will be no way to tell how young it is – or if it's a small area, whether its lack of craters is due to its size or its age. Then there's the effect of an atmosphere: the thicker the atmosphere, the larger a meteorite has to be to keep enough of its interplanetary speed to make a crater when it strikes the ground. And as we'll see, the thickness of the Martian atmosphere has changed a very great deal, more than enough to affect our calculations.

But the biggest problem with crater counting is that while we know that the rate of cratering is an uneven process, we don't know how uneven it is. Look in the bucket of rocks we've been shovelling up and throwing down. We can see that in our first few throws we took almost all of the bigger rocks, in the next few we were left with several middling-sized pieces and some gravel, and now we're scratting around at the bottom to get anything. Going backwards from our analogy, this is precisely what happened in the early solar system.

We just don't know the make-up of the debris swarm that intersected Mars's orbit early on. But we do know that, like the rocks in our bucket, meteorites can only ever strike an object once. They're not a renewable resource. Shortly after formation, Mars was subjected to an intense bombardment of leftover space debris, some of which included kilometre-wide planetesimals. After these were cleared by collisions with planets, both the number of impacts per year and the size of the potential impactors decreased rapidly.

The biggest aid to dating the Martian surface comes not from Mars itself, but from the Moon. The two aren't particularly alike, and they formed in different parts of the solar system by different mechanisms, but one vital similarity links them: their craters. Crucially, we have some samples of the Moon, which we can not only date using radiometric dating, but because we know exactly where they came from, we can compare landscapes. With a few adjustments, we can devise a scheme to transfer the whole timescale of cratering rates on the Moon a hundred million kilometres away to Mars, and roughly – sometimes very roughly – date the various ages of its surface. Nothing trumps walking the landscape, examining the structures close up, taking samples and drawing on a paper map with coloured pencils. But these methods are the best we have for now, and they work far better than they have any right to.

Mars's landscape might appear alien at first, but by spending time looking at the folds of the land – the craters, ridges and valleys – we start to appreciate that the processes that sculpted Mars are explicable – each age dominated by impacts, water, fire and ice. Reading the shape of Mars gives us the words to describe both its past and its present.

MARTIAN METEORITES

Much of the purpose of sending sophisticated probes, whether they are intended to stay in orbit or to land, is to take our laboratory to Mars. Spaceborne lasers can measure not just the height of the land but also the composition of the air, exposed rocks and ice. Photography has revealed weather patterns, dust storms, the seasonal variations in landforms, the growth and retreat of the ice caps and glaciers, the movement of sand dunes and tantalising hints of surface water. Magnetometers detect the residual magnetic fields imprinted on rock formations.

Down on the planet, rovers have tasted the air, drilled the rock and driven kilometres across the surface in search of new vistas. They've carried weather stations to record the daily rise and fall of temperature, pressure and wind speed. Their cameras have looked up at Mars's moons transiting the sky and the pale-yellow disc of the Sun. They've even taken pictures of the faint dot of our own planet. But we've been unable to send anything that might prepare and then analyse a sample for radiometric dating. The equipment is simply too big and energy-intensive to take.

The plan is to ultimately gather up a bag of carefully labelled rocks from the Martian surface and carry them back to Earth – the newly arrived Perseverance rover will collect samples, a second rover will load them on a small rocket, and a third mission will retrieve them from Mars orbit, sometime in the 2030s. Back on Earth, we can subject the samples to the full panoply of scientific tests. We'll have the advantage of knowing exactly where the samples came from and other contextual information about them: how weathered they are, what types of rocks were above and below them, whether they show signs of alteration or impact shock, whether they were from a crater wall or floor or from a rock bed outside, whether they were taken from a contiguous formation or a broken block of ejecta, and even their north, south, east and west orientation. All of this will help us to interpret the raw physical data.

But even without that extra information, any verifiable piece of Mars is going to be fantastically valuable for research. We've already discussed how a very large impact on Mars can throw material into orbit. Beyond that, it is only a short step to realising that some of that Martian ejecta might have already found its way, through the clockwork of orbital mechanics, to other planets and landed as meteorites there.

Identifying particular meteorites as having come from Mars has been a difficult process and it was only recently settled using information from the Martian landers. What was known beforehand was that some rare meteorites weren't like any of the others. The majority of meteorites are chondritic meteorites: that is, mainly composed of chondrules. The majority of the others are from broken planetesimals which, like Mars, have either completely or partially melted: these are the meteorites which are all iron, mixtures of iron and rock, or now-frozen magma.

The Martian meteorites were originally assumed to be from

the last of these groups, but they had unusual chemical characteristics that set them apart. The iron minerals were oxidised. Some were affected by recent low-temperature alteration with carbonates, salts and clays. The magmas from which they formed showed the presence of water. The mineral crystals in the frozen magma itself were different to those found in other silicate meteorites. When analysed for their radiometric ages, they didn't all home in on 4.6 billion years, like almost every other meteorite. Some were very old, but still not that old, and some were considerably younger.

Crucially, there were minute traces of gases trapped within the rock, in crystals that were decoded to be shock-glass, which is created by instantaneous heating as a pressure shockwave passes through the rock. It was realised that these gases could only be pushed into the glass if, at the moment of melting, it had been exposed to air. After the Viking lander provided a basic analysis of the current Martian atmosphere, the similarities between this and the gases trapped within the meteorites were too great to ignore. Every difference between an asteroid-derived meteorite and these other, stranger ones could be accounted for by proposing that they came from Mars.

Do we find meteorites that come from the Moon too? Yes, and the chemical signatures of these are identical to the samples retrieved from the Moon by the Apollo missions – in fact, their recognition smoothed the way for accepting that unaltered, unvaporised debris from a meteorite impact on another celestial body can end up on Earth. Once we get curated samples back from the Martian surface, bagged and tagged, any lingering doubts will be laid to rest, but for now it's accepted that these meteorites – numbering 266 out of more than 60,000 known space-rocks – are examples of Mars.

These rocks tell us a great deal, but they also come without context. They are random pages torn from different books, and while we can tell where in their stories they come from, we don't know what happened before or after, or where that particular passage was written. They are all igneous rocks of some kind – either lavas that flowed on the surface or deeper emplacements of magma that were later excavated. All have recorded more than one stage of alteration, though, before the final event that saw them launched into space, to fall to Earth, where they were collected by us.

These few rocks give us a benchmark of chemical compositions, some scattered ages and tiny insights into what was happening on Mars at the times they were formed. One particular Martian meteorite is ancient, at 4.5 billion years old. But some of the meteorites appear very young, around 180 million years old, and most troublingly for our understanding of the planet, they suggest both that Mars was geologically active far later than previously thought, and that free water was present on the surface well beyond the time when it was presumed to have vanished. Like ghosts, these two ideas will haunt the rest of this story. Whenever we're tempted to say that Mars has finally fallen asleep, we can hear these rocks whispering, *not yet, not yet.*

THE GREAT DICHOTOMY

D ividing the history of Mars into stages is largely a matter of
established convention, but our first major reference point
is going to be the formation of the Hellas crater. Its creation
resurfaced most of Mars with its planet-spanning ejecta blanket – the evidence for almost everything that happened earlier
was buried.

Parts of the Martian crust resemble the most heavily cratered surfaces on the Moon. Typical of these earliest surviving
post-Hellas landscapes is Noachis Terra, a section of the southern highlands, and it's from there that we take the names of
the oldest Martian eras: the pre-Noachian, to denote everything
that happened before Hellas, and the Noachian, for events that
happened afterwards. The date we put on the Hellas event is
somewhere around the four-billion-year mark, but there's some
debate about that – if we call it 4.1 billion, we'll be roughly right.
It's older than any rock formation on Earth, for certain.

Conditions on pre-Noachian Mars are difficult to ascertain, if
not unknowable. Most of the information we might want to look
at has been destroyed, reworked or covered over. We can deduce
little about the planet's atmosphere, its surface conditions or its
major landforms, and we can only tentatively discuss its internal
processes, the state of the core and mantle or how vigorous its
volcanic activity was. Most likely, though, pre-Noachian Mars

was a place of almost unimaginable horror, a genuine representation of what we might think of as Hell.

Despite the vigour of that impossibly distant and most ancient time, evidence of events then has managed to survive until now – chief of which is Mars's greatest and most implausible feature. When we were finally able to draw an accurate altitude map of Mars, it showed baldly what had been partially known for years: the northern half of the planet is significantly lower than the southern half. This is known simply as the Great Dichotomy.

It's something that makes Mars different. Not just slightly different, but wildly, strangely different. In detail: the surface of Mars is at least 2.5 kilometres lower in the north than it is in the south, and there are parts where the difference is greater than 6 kilometres – this is an actual vertical difference in height. The junction between the two regions is, for at least half its circumference, marked by an obvious slope, taking the elevation from the north to the south up to between 2.5 and 3.5 kilometres. It's not a cliff face; instead it's a gentle but relentless ramp, gaining height along a horizontal travel of hundreds of kilometres.

Over that distance the landscape changes. To the south, the land is high and characterful. Craters are everywhere, crater upon crater upon crater, subsequent impacts having overwritten the record of earlier ones. Ejecta blankets and lava flows interleave each other. Using the technique of crater counting, we can judge the exposed surface of the southern highlands to be an enormous age: it measures as Noachian, but it's hammered into the same primal pre-Noachian crust that Hellas struck.

By contrast, the north – dominated by sparsely cratered flat plains – is cloaked with deposits that are a billion years younger than those in the south. Deep beneath the plains lie the bones of huge, ancient, eroded impact craters that manifest as quasi-circular depressions. Only their vastness thwarts their complete burial.

The Great Dichotomy is far from skin-deep, though: it's mirroring something that's happening deep underground at the base of the crust, where it meets the mantle. We've been able to study this thanks to the same satellite that gave us the altitude data – it contributed to another, equally important map: a gravity map.

From first principles, gravity is a result of mass. All physical objects – from the largest supergalactic structures to the tiniest flecks of dirt – have mass, and therefore they all exert a gravitational force. The more mass an object has, the greater its gravitational effect, and the movements of objects are determined by the interactions between them: moons orbit planets, which orbit stars, which orbit the central galactic core. Also, the closer objects are to each other, the more influence they have; gravity is four times stronger at half the distance. Having said that, gravity is an incredibly weak force. Just think of a person standing on Earth: we usually manage to hold our own bodies upright against the pulling power of a billion trillion tonnes of rock.

Gravity is almost, but not quite, the same value across the entirety of Mars. It takes very sensitive instruments to detect the parts-in-a-million differences, and we can also very closely monitor the orbits of satellites around Mars to see if they deviate from their predicted paths. Where gravity is weaker on the planet below, the satellite will rise ever so slightly as it passes. Where it is stronger, the satellite will dip down. The difference is in millimetres, but we can measure it.

Large masses of rock, like mountains and volcanoes, create minute amounts of extra gravity. Likewise, huge craters have a missing mass and there is less gravity. A first-order map of gravity – called 'free air' gravity – will often resemble the relief map, where mountains have positive values and craters negative ones. But the power of a gravity map is its ability to dive under the superficial and measure what is happening below. The gravity values can be adjusted for the visible discrepancies in height and a second map emerges. It tells us where the gravity of Mars deviates from what we expect, and among the things it tells us is this: the Great Dichotomy is baked into the structure of Mars.

The average thickness of the light silicate crust that overlies the denser mantle varies as we move from north to south. In the north, it's an average of 32 kilometres thick. In the south, it's almost twice that, at 58 kilometres. The distinct difference between the two regions means we can draw a line all the way around Mars, even where the visual signs have been obscured on the surface above.

How did this startling feature, unique among all the planets, come about? We have two main explanations and little to choose between them. Each one has its supporters. Each one has evidence to back it up. Rather than choose which to believe and which to reject, I'm simply going to present both. They can't both be true. But because they both end up at the same place – with the Great Dichotomy girdling Mars – there's no harm done. We can entertain both ideas without committing to just one.

THE GREAT DICHOTOMY
CONVECTION THEORY

Back when Mars was molten, the iron–nickel component sank to the centre to form the core, and the lighter silicate crust rose to the top. Between them were the denser, heavier silicates of the mantle which, while hot enough to be molten, were under so much pressure that they behaved more like slow-moving putty than a liquid. The crust cooled, forming a rigid lid, but underneath, everything else was still hot and continued to generate heat.

Heat is transferred in three ways. First, radiation: what you feel when you put your hand up to the Sun is the direct effect of radiation – the energy from the light striking your skin. Second, conduction: this is what you feel when you put your hand on a radiator and touch the hot metal, which is heated by the water inside it. Third, convection: this is what you feel when you put your hand above the radiator, as the air heated by the hot metal rises to mix into the air in the room.

All things above absolute zero (–273°C) radiate heat in the form of light, and the hotter they are, the more radiation they produce. The energy of the light depends on its wavelength, and we can measure that to find out how hot an object is. At cooler temperatures, the light is infrared – invisible and long-wavelength – but as a material heats up, it begins to shine in visible light, which has a shorter wavelength. A hot object will first glow dull red, then bright yellow and finally blinding white. The temperature of the material dictates the colour and the intensity of the light coming from it.

Conduction is simply the ability of a material to transfer heat energy from the hot side of itself to the cool side. Metals are very good at this; rock, not so much. Liquids and gases are terrible at conduction, since when they get hot, they expand and, being less dense, they rise up, taking their heat energy with them.

Which brings us to the third mechanism: convection. Convection is why your radiator is warmer at the top than the bottom: inside the radiator, the hottest water is the least dense and it rises as far as it can go. When it can't rise any further, the trapped water passes over some of its heat energy to the metal of the radiator through conduction and the air by radiation. This cools the water down and it sinks, to be replaced by more rising hot water in a continuous process of circulation.

Convection also applies to hot, plastic mantle rock. Held in a semi-solid state by the overlying pressure, it can still move very slowly. Heat at the bottom of a thousand-kilometre column of mantle rock is initially trapped, but if the rock starts to flow, both the rock and the heat it carries can be transported upwards at the rate of centimetres a year. Given a long enough time span, that rock will arrive at the bottom of the crust, where the heat will be conducted imperfectly away, up and out towards space, and the cooler rock will sink back down again in a huge, slowly stirring gyre.

The greater the difference in temperature between the surface and the core, the faster the mantle moves from deep down near the core to up against the base of the crust. And like a deep and secret current, the mantle starts to push against the crust, dragging it along as it sheds its heat and then plunging down into the depths again.

This is the exact mechanism that is happening to this day on Earth. Plumes of rising, hot mantle, starting at the core–mantle boundary and ending below the crust, drag plates of crust with them as they turn over and descend. The surface plates pull apart from each other in some places – where we find lowlands, basins, seas and oceans – and collide edge-on in others; where collisions happen, one plate rides over the top of the other – forming mountain ranges on one while the other is forced underneath. Each fracture line between plates is associated with volcanoes and earthquakes, and the motion is constant, gradual and glacial – centimetres a year, at most.

But did this ever happen on Mars? It's difficult to say, and the answer is probably yes and no. There was certainly a failure to sustain planet-wide mantle convection and a proper system of plate tectonics. If it ever happened at all – and the evidence is scant – then it died away quickly. That might have been due to Mars's size or the material it was made from (or both, or other factors we haven't divined), even though the core was still belching out heat and the mantle was hot and soft.

What we do have evidence for, though, is one very large plume – and this is the basis of one theory for the Great Dichotomy. If there was a singular plume of hot mantle material being driven upwards by heat from the core, directed roughly towards what is now the south pole and pushing against the crust until it cycled down again, cool, at the north pole, then we'd get exactly what we see: a high, thick southern highland, where extra material has been pressed onto the base of the crust, and a low, thin northern plain, where the base of the crust has been peeled away, dragged down and entrained in the sinking mantle below.

Computer modelling can show that this concept isn't just a fever dream, and that a planet of Mars's size and presumed composition could operate just one plume to the exclusion of others. In fact, as the single plume began to dominate, it would have become so efficient at removing heat from the core that any other plumes that might have challenged it would have faded away.

The most immediate and compelling evidence that suggests this may have happened has to do with the behaviour of Mars's core, which is one of any planet's least knowable parts. We can never sample it, never see it and can only infer its presence and composition by indirect measurements. Despite this, we know that Mars has to have a core, on the basis of how big and how heavy the planet is: we can calculate how much of Mars is rock and how much is metal.

The cores of rocky planets – ones made of predominantly iron and nickel – start off as fully liquid, intensely hot but trapped in the centre, crushed by the weight of rock all around. A solid core starts to form soon after at the very middle. This fraction of the core slowly increases as heat escapes through the mantle and crust, until the once-liquid core solidifies completely.

But in the between-times, when there are both solid and liquid metal in the core, the liquid will move in swirling convection currents, just as the mantle above it does but vastly quicker.

Those currents are affected by the rotation of the planet, and they all spin in the same direction.

Liquid metal conducts electricity just as solid metal does, and if that conducting fluid moves through a magnetic field – like the one flowing out from the Sun – it produces electricity. Those electric currents in turn create their own secondary magnetic field, which goes on to support the generation of electricity, and so on. A self-sustaining electromagnetic dynamo switches itself on like a light bulb, powered by the rotation of the planet, and it will last as long as the conditions are right – when there is both rotation and a liquid core.

Back at the Martian crust–mantle boundary, a thousand kilometres above the core, the plume of hot mantle added kilometres of rock onto the underside of the crust by injecting semi-molten magma into the lower levels. It simultaneously dragged the crust along as the hot mantle turned ninety degrees and scraped northward along the underside of the crust. Magnetic minerals within the crust – at this point, likely to be in the form of the iron oxide mineral magnetite – were reheated by the new material, and when they cooled again, they preserved a record of the direction and strength of the magnetic field produced by the dynamo in the core, writing the information into the rock like a piece of magnetic tape.

One of the things we know about planetary dynamos is that they flip the direction of north and south at various time intervals. If the magnetic tape of the rock continually records the passage of time as it moves, then it inadvertently stores the north-then-south-then-north-again direction of the magnetic field as a series of stripes. And we can see those stripes from orbit, using a satellite equipped with an instrument that measures the magnetism of the planet below. Although the dynamo no longer works in the present day, the evidence showing that it once did appears as a pattern of differently magnetised bands centred around the south pole like a bull's eye. Imperfect, battered by later crater-forming and volcanic events, but tantalisingly present all the same.

This left us with what we see on Mars now: half a planet with a thick crust, centred close to the current south pole, that was pushed and squeezed northwards as a creeping tide. The thinner crust was nudged ahead of it and thinned by the sinking mantle near the north pole. But before we take this on trust, we have to consider the other scenario.

THE GREAT DICHOTOMY
IMPACT THEORY

If you find the convection theory a bit of a stretch, how about this? The Great Dichotomy resulted from a single, giant collision that left an elliptical impact crater roughly the size of the entire northern hemisphere.

For this theory to work, we're dependent on the serendipitous coming-together of several factors entirely external to Mars. First, we need a giant impactor, a leftover planetary-embryo-sized one, some 2,000 kilometres across. Secondly, we need it to be moving slowly relative to Mars: somewhere around a 6–10 kilometres per second difference (this is slow in cosmic terms). Thirdly, it has to strike Mars's northern hemisphere at an angle of around sixty degrees to the vertical.

Conditions outside of these won't give the results we want. A smaller impactor would leave a regular, albeit huge, crater. A larger one would remelt the entire surface of Mars. A slower impact would add extra material to Mars, not take it away, but a faster impact would strip Mars of much of its mass. A head-on collision would produce a circular shockwave that would travel around the planet and meet at the back, 180 degrees from the impact site, blowing a smaller but still significant hole through the crust there too. A shallower-angled impact wouldn't give the shape to the northern lowlands that we see.

We've already accepted that 'spare' planetary embryos existed: that Mercury is almost all core and no mantle is likely an example of a planetary embryo impact, and the formation of Earth's Moon is strongly believed to have been the result of a Mars-sized embryo colliding with the proto-Earth, 4.5 billion years ago. No physical evidence of either of these impacts remains: they were simply too large and too energetic.

Slow-moving planetary embryos sharing orbits is also not too much of a reach. There are stable points – Lagrange points – one-third of the way around a planet's orbit, both ahead and behind it, where a decent-sized planetesimal might accumulate material without falling into the main planet. These orbits are stable over hundreds of thousands of years, but we're operating in time spans of millions of years. Any perturbations caused by, say, wandering gas giants moving into the inner solar system and then back out again would disturb the delicate balance of gravity between the Sun and Mars, and anything at a Lagrange point might be nudged away from it. Because such an embryo shared the same primary orbit around the Sun, Mars would eventually capture it – not in a full orbital-velocity-plus slam, but at a slower, more considered shunt.

Oblique-angle impacts are obviously possible: there's no law of nature that insists an incoming meteorite has to hit at right angles to the ground. But how does an impactor's angle of approach affect the shape of the crater it leaves? We've already learned that it's not the actual downward movement of the impactor that excavates the crater – it's the explosive expansion of the superheated vapour created from both the impactor and the surface directly beneath it. That should mean that all impact craters are circular. And they are . . . almost. If the initial fireball is elliptical, because the impactor is large enough to have smeared itself across the target before it vaporised, then the resulting crater will be longer in the direction of travel than it is wide.

There's a sweet spot, then – a Goldilocks scenario that is both entirely possible and highly improbable. It doesn't matter how unlikely it is, though, because what we see in Mars's northern hemisphere can be explained by this single, huge impact. Something caused it, so why not this?

In the moments after the collision, an elliptical crater formed, covering almost the entire northern hemisphere. A supersonic shockwave of heat rushed outwards, eventually encompassing the whole globe. Debris from both the impactor and Mars followed as a near-solid wall of material. Some of it was launched into space – as incandescent gas, cold broken rock and every state between. The crater, deep and very broad, filled with a deep sea of lava. Much of the atmosphere of Mars – the dense, steaming, thick, early atmosphere – was blown off the planet and lost forever.

Then the curtain of excavated material – that which hadn't achieved orbital velocities – started to descend. It was literally a rain of fire from above, one that resurfaced the rest of the planet. A larger, more energetic impact would have melted the planet for a second time, but that didn't quite happen. Still, the effect was global and complete. Much of the early crust in the north was added to the broken south in a tens-of-kilometres-thick ejecta blanket.

As it cooled, the lava lake settled into the crater floor, producing flatter, lower terrain. Much of the crust had been deeply cracked, not just underneath the crater, but everywhere, from top to bottom. The northern mantle relaxed and rose after much of the weight of the crust was lifted off it. Conversely, in the south, the extra weight pressed down, causing the mantle to heat up. The rim of the crater, initially well defined, slipped and slumped along fault lines as the deep rock underneath moved to accommodate the new situation. As the debris in orbit continued to spiral back down, it cratered and marked the new landscape below.

Mars was left with an abnormally low northern hemisphere and an abnormally high southern one, but the age of the top of the crust beneath each dates to the same period: one was formed by the cooling of impact melt, the other was formed by the shovelling of a wall of hot rock out of one hemisphere and onto the other. In an age where whole protoplanets fell into the Sun and others were torn apart and remade, it seems remarkable that we might still have a record of an event that almost, but not quite, broke Mars.

Two stories, then. Either might be true. Both might be false. And there are other explanations, one of which even suggests that the first theory could have resulted from the second – a giant impact set off mantle convection by providing a thick insulating southern cap, which built up heat beneath it and encouraged the reluctant mantle to move. We're in the dark here. More information might settle the matter, or it might confuse the theories that we already have.

However it formed, we are left with the Great Dichotomy. Nothing quite like it is seen anywhere else in our solar system. Among other merely spectacular features, it's unique to Mars.

It's so significant that it dominated the way the planet developed over the next 4.5 billion years. The immediate effect was to leave Mars wildly unbalanced forever. Wherever the initial impact or plume occurred, the difference in crustal thickness meant that the whole planet tipped over to a more stable configuration, leaving the Dichotomy lying around the equator.

The ancient Noachian landscape, with its record of the earliest impacts, now lies completely buried in the north, under later deposits: all that we can detect of it are the deep wounds that manifest as quasi-circular depressions, hidden under layers of lavas and inflowing erosion products from the highlands – sandstones and mudstones and salts. In what became the south, the primordial surface remains exposed, exhibiting its elementally inflicted scars for all to see.

PHOBOS AND DEIMOS

Just as a planet orbits a sun, a moon orbits a planet, bound to it by ties of gravity. Moons can vary enormously, in size, composition and shape. Some, like Ganymede, a moon of Jupiter, and Titan, a moon of Saturn, are worlds in their own right, and only Mercury and Venus do without moons in our solar system. Even Pluto, recently demoted from planet status, has a significant moon called Charon. That Mars has two moons, Phobos and Deimos – named after the twin sons of Ares, the Greek version of the Roman god Mars – is blessedly unremarkable.

This being Mars, though, nothing is ever simple. While these moons are somewhat potato-shaped and very much on the small side – Phobos's longest dimension is 27 kilometres, Deimos's is 15 – this is not specifically troubling. The first odd thing is that Phobos's orbit is really very close to Mars, a mere 6,000 kilometres above the surface. This is so close that it can't be seen from the surface above seventy degrees north or south – the moon is always below the horizon towards the polar regions. Phobos is so close that it orbits faster than Mars rotates: rather than rising in the east, it rises in the west and apparently tracks backwards across the sky, twice in one day. It moves quickly enough that, if we were to stand on Mars and look up, we'd see it moving, west to east, in a little over four hours. It's so close that it visibly grows as it passes overhead, then shrinks into the distance.

Yet it'll move closer still. The same gravitational forces that have locked Phobos into only ever showing one side to Mars are also causing it to gradually fall towards the planet at a rate of just under two centimetres a year. Not that it'll ever strike the surface – before then, Mars's gravity will tear Phobos apart and smear it across the heavens. What is now a lumpy but coherent piece of rock will become a long train of debris in some ten million years' time, which will then spread out further and form a planetary ring. The falling-apart of Phobos will take little more than a month.

Part of the reason for Phobos's future disintegration is that it is barely there. If it was made from solid rock, its density and surface gravity would be twice what they are. Instead, Phobos seems to contain a lot of nothing. Depending on what we believe Phobos to be made from, the amount of nothing varies from around one-fifth to nearly half its volume. But since we only have ideas as to the composition of the moon, the estimates remain very much in the realm of guesswork.

It's suspected that Phobos is made out of some sort of accumulated leftover debris. It's uncertain whether this came from a small, captured rubble-pile asteroid or a small agglomeration of planetary debris thrown up from the surface of Mars by a giant impact early in its history – like the one that may or may not have caused the Great Dichotomy. There are other candidate craters we can see, and potentially ones we cannot.

We need to pause here to consider one of the more bizarre but of-its-time-rational theories: that Phobos is an alien construct, and that its low density is explained by it being nothing more than a hollow metal shell. Russian astrophysicist Iosif

Samuilovich Shklovsky seriously proposed this in 1958 – that given its density and size, the idea that Phobos is an iron hull around fifteen centimetres thick surrounding a vast interior space adequately explained everything.

It took a good decade before anyone could prove him wrong, and we know for certain now, having better photographs of Phobos, that it isn't a relic of a Martian – or anyone else's – empire. We do know that Phobos is extraordinarily dark, though. Albedo is a measure of reflected light. Fresh snow reflects around 90 per cent of light; Phobos reflects just 7 per cent. The whole of Phobos is covered with a deep, hundred-metre layer of broken rock and dust. It's almost perfectly black, except that there's a reddish tinge to its tail and a bluer tinge to its head.

Also at the leading edge of Phobos is Stickney crater, caused by an impact that must have been close to the limit for breaking apart the entire moon. Stickney is 9 kilometres across, on an end-on diameter of only 20 kilometres. At first glance, there are lines, ridges and faults radiating out of Stickney across most of the surface of Phobos, but the lines also cut Stickney's crater rim and no one is certain what they are or how they got there.

Stickney is not the only crater, though, and now that Phobos has been mapped, an attempt at crater counting – and therefore giving it an age – can be made. If Phobos formed around Mars, then it's very old: 4.3 billion years old. If it didn't, and it was captured from elsewhere in the solar system, where impact speeds were a third less and the craters made were comparatively smaller, then it would be a mere 3.5 billion years old.

Deimos, the smaller, more sensible brother, is further away from Mars at 20,000 kilometres distant, and it's moving away from the planet because it sits on the other side of a gravitational line: in 100 million years it will slip Mars's orbit completely. It does, at least, have the decency to rise in the east and set in the west, and its orbit takes only slightly longer than a single Mars day – it takes nearly two and a half sols for Deimos to fully cross the sky.

Deimos is made exclusively of the red dirt found on Phobos, but it's even less dense than its brother. It's marginally darker too, and smoother. The crushed-rock surface of Deimos might be in the region of a hundred metres thick, like on Phobos, but it's barely held in place. Because Deimos is so very small and has so little gravity, crater-making impacts will throw almost all their ejecta into space. From the few photographs we have of Deimos, it might just hold the record for the largest crater in relation to the size of the object. The unnamed hollow at the south pole is 10 kilometres across, on a surface that is only 11 kilometres wide at that point.

From the ground, and despite its darkness, its proximity makes Deimos appear as one of the brighter stars in the sky. But neither Martian moon has ever troubled events on the planet's surface. A Martian ocean has never risen up to greet them. The Martian ground has never flexed beneath them. They are too small, too insignificant to affect the surface. But they are there, and we acknowledge them all the same.

THE EARLY MARTIAN ATMOSPHERE

All of the ices, and all of the gases caught up in them, boiled when Mars melted. Just how much ice and gas ended up as part of early Mars is another question for educated guesswork to answer, but it would have been a lot, and potentially a very great deal indeed.

Mars formed from debris that was close to the snow line around the proto-Sun, at the point where water turned from a gas directly into a solid through the process of deposition. A rough calculation that involves making Mars out of comet-like planetesimals consisting of half water and half rock leads to an estimate of well over a hundred billion tonnes of water included in the material that formed Mars. If all that water was liquid, it would cover a Mars-sized sphere to a depth of between one and two kilometres.

But given the temperature of early Mars – a still-hot, semi-molten planet skinned by a fractured, congealing crust, riven with eruptions of lava and instant heating caused by repeated, giant meteorite impacts – the surface at that time was too hot for liquid water. All of the ocean's worth of water was steam, and when combined with the carbon dioxide that had streamed out of the rock along with it, Mars most likely had an atmospheric pressure shortly after formation somewhere between 100 and 300 bars. That's the same pressure as experienced between one and three kilometres down in Earth's oceans. Today, Mars's average air pressure is 30,000 times less.

This was a searing, crushing, highly chemically active atmosphere. The temperature fluctuated wildly. When it fell low enough for water to condense out, it would have rained in torrents: temporary lakes, seas and even oceans would have gathered in low areas, with the water draining through cracks in the crust caused by impacts. Here it would have met the intense underground heat and re-emerged as steam: geysers, hydrothermal vents, pools of superheated boiling water under a dense, hot water-and-carbon-dioxide atmosphere. When the temperature rose again – after a large impact – those seas would have boiled off, raising the air pressure even higher. The greenhouse effect – the ability of the atmosphere to absorb and retain heat – would have played a significant part in keeping the surface of Mars hot, too, as both water and carbon dioxide are efficient greenhouse gases.

And yet that early, thick, water-heavy atmosphere is unequivocally gone now, stripped away and replaced by something almost entirely different. Where did it go? It seems to have vanished as quickly as it formed, in a few tens of millions of years, and we have to hunt for potential mechanisms that might have contributed to this act of atmospheric theft.

Giant meteorite impacts can accelerate the column of air above them, punching it into space, never to return. But we also know that meteorites are a net contributor of volatile gases and ices to a planet: if we were to bombard Mars with comets, no matter the wreck they made of the air that was already there, the fact that the comets brought ices to the planet means that we'd ultimately end up with more atmosphere than we started with, not less. So while atmosphere loss from meteorite impact was absolutely a factor, the extent of it depends on what the meteorite was made of. Early in the life of Mars, this doesn't look like the culprit.

The atmosphere could have simply wandered off the planet. Mars has a low surface gravity, making the column of air above Mars very tall and the escape velocity – the speed at which something needs to be travelling to leave Mars's orbit – proportionally low. At the very outer reaches of the atmosphere, it only takes a nudge for air molecules just to drift away into space.

The Sun continuously emits a stream of charged particles from its yellow-hot surface, known as the solar wind. These broken parts of atomic matter can strip gas from a planet by colliding with the top of the atmosphere at high speed and, like billiard balls, knocking the gas away. Planets with a dynamo-type magnetic field deflect those solar particles around the atmospheric envelope, while those without suffer dramatic losses of air. But early Mars had such a magnetic field – it was protected. When it failed, solar wind predation became important, but for now, it's not in the frame.

The most likely answer is that Mars's atmosphere was inherently unstable because of the planet's low gravity. Mars's early hot and energetic atmosphere ballooned to not just hundreds of kilometres above the surface, but thousands, and possibly as far as twice the planet's radius. This extraordinarily massive gas envelope around a small rock core would still have been protected from the solar wind particles by Mars's magnetic field, but it would have sat in the full, unflinching gaze of the Sun just as it began to settle into its hydrogen-burning phase. The star's output of short-wavelength ultraviolet light and X-rays provided more than enough energy to split molecules and send individual atoms over the escape velocity barrier and away. Any gas that found itself at the distant margins of the atmosphere escaped easily into space.

It was just a question of when – not if – Mars's giant atmosphere would dwindle and fade. From being over 100 bars at its peak, the combination of low gravity, high heat and vigorous sunlight pushed the atmosphere out and away. Some 100–150 million years later, the surface pressure was down to perhaps one or two bars. By that time, almost all the icy planetesimals were gone, having already collided with other objects or settled into stable orbits elsewhere. The only source of fresh atmosphere available was what Mars had managed to hold on to inside itself: carbon dioxide, sulphur dioxide and, critically, water, kept within the crust, where it had percolated down through cracks and into aquifers – great storehouses of liquid below the surface.

The Great Dichotomy had already formed. The northern lands were kilometres lower than the south. The pressure dropped as most of the atmosphere bled away, and with it the surface temperature. The cooling water vapour and carbon dioxide atmosphere wasn't stable and would, over the next four billion years, diminish to almost nothing. But the conditions at that moment were right: if water was ever going to flow freely on Mars, it was in the early Noachian age.

PART FOUR

THE NOACHIAN

(4.1–3.7 billion years
before the present)

SAILING ON AN ENDLESS SEA

This is much more like the adventure you were promised. You are about to sail on the ocean of Mars – or you would if you had a sail, but this is the next best thing: a little inflatable boat equipped with a small, robust and necessarily enclosed electric motor to move you about.

At the shore, the waves behave strangely; the one-third gravity makes them tall and thin. It's difficult to launch the boat in these conditions, as the white horses try to drive the hull back onto the beach. The wash pours in and drags out across the rocky beach, booming and roaring. You wrestle the boat through the surf with help, and a spacesuit that's totally sealed against vacuum is good enough for the shallows. The motor works perfectly and you dip the fast-turning propeller into the water – you're off.

Every time you skip over a wave, you feel like you're airborne for a moment: the propeller spins wildly as it claws at the air, then gives you a surge of acceleration as it submerges again. You finally escape from the line of breaking waves and reach deeper water, where the swell is still severe but manageable. No one wants to throw up in a spacesuit.

You have been given the task by virtue of your summers by the sea. You know how to handle small craft, but this is the strangest shore you've ever departed from. The land is a dense black line. The rock is dark. The sand is black. There's little variation in colour: there's no oxygen in the air to react with the minerals and turn them red, or any other colour. The land slopes down inexorably from the south to the north; any visible change in height is either due to an eroding crater wall or the scar of a river.

You call them rivers, but they're more like wadis – dry river-beds that burst into short-lived exuberance after a storm has passed inland. There's nothing to slow the run-off, no soil, no vegetation to soak up or impede the flow of water from the land back to the ocean. Every time it rains, the land is scraped clean and the rivers rise in their gorges so fast that a wall of water charges downhill, bearing everything from fine silt to boulders, battering and cracking the banks and the beds, and then subsiding almost as quickly as it appeared. A pour, a trickle and stop.

And now you're afloat on that reservoir of water. There are lakes in the highlands, where huge craters have warped the surface and produced deep local lows that will never be overtopped. They have their own weather, but you're on the great northern ocean and there's nothing but waves between you and the pole. In fact, as there's no ice, you could – in theory – cross that point and head south again, to land a full 180 degrees of longitude from where you started, a journey of almost 10,000 kilometres. Wider than the Atlantic. Almost a Pacific crossing.

It's why the swell is so high. The winds rattle around the hemisphere of water with nothing to deflect them. They're always dragging on the surface, always raising waves. Sometimes there's a storm – there's always a storm, it just depends where – which gathers itself up and bulls its way inland, releasing torrents of water onto the north-facing Dichotomy rise: as the rain clouds are pushed by the slope into higher, colder air, every last drop is wrung out. The southern high latitudes are almost a desert.

And sometimes, there's a bigger wave still. With half a planet of ocean, a meteorite strike can produce a tsunami of gargantuan proportions. This colossal wave can smash against the shore, tearing it up, flinging debris up the slope and boring deep along the same river valleys that carry the water down. Some of the waves you're skipping along the top of are echoes of those events, passing and repassing around the circular sea.

While you still have an important scientific mission to carry out, you can't help but feel the joy of zipping off into the distance with your load of instruments and sample containers. You reach the first waymark and cut the motor. Your sampling device is no more complicated than a weighted length of line with a bottle strapped to it. When you press the trigger at your end, the bottle will open and suck in the water, and it will close again when you let go. There are also temperature gauges and a device to log your position for that sampling event. The most difficult thing you have to remember is to use the numbered bottles in the right order.

As you go further out, the shoreline sinks rapidly. If it was perfectly calm, you'd lose sight of it after a mere five kilometres. But it's never calm, so the distant dark line pops up as you crest a wave. As it passes, it blocks your view until the next time. You're surrounded by a bowl of water, with only the cloud-studded sky above. You can imagine being out here with no idea of where land was, and instead of turning south, going in any other direction and becoming hopelessly, wildly lost. Except you're a better navigator than that. The Sun might be confusingly high now, but it still sets in the west and you know south is to its left.

You restart the motor and plough on through the waves to the next sampling point. You could almost forget where you are, and be tempted to unseal your suit and let the salt water slap against your bare skin. If you held your breath, you might even survive the experience. There's no one out here with you, either to stop you or to help if you get into difficulties. It's just too risky; the moment passes. You'll always remember it, though, and wonder if you shouldn't have been a little more reckless.

THE HELLAS IMPACT

Hellas is the single largest extant crater in the entire solar system, at 2,300 kilometres across and 9 kilometres deep (but it's only the second largest impact feature on Mars, after the enormously eroded Utopia Planitia basin, or the third if we credit the Great Dichotomy impact theory). Its creation marks the end of the pre-Noachian and the beginning of the Noachian. In a history which already contained repeated huge impacts of global significance, Hellas was still an absolute monster. The ejecta was thrown pole to pole, enough to bury the entire surface of Mars half a kilometre deep; the temperature soared and the air pressure rose dramatically as liquid water turned back into steam. The immediate effects lasted for decades and centuries, but others persisted for far longer.

The amount of energy released by such an impact is almost impossible to comprehend. As well as excavating some seventy million cubic kilometres of Martian rock and launching it up to halfway around the planet, and even into space, much of the violence of the explosion went down into the crust. Cold rock is brittle. It's strong when compressed, under a heavy load bearing down from above, but it's extraordinarily weak in tension. Breaking rock is easy when it's simultaneously being pulled apart.

In the moments after impact, when the pressure wave was driving through the crust and into the mantle, everything was in compression. After the explosion, all that pressure abruptly disappeared: the energy stored in the rock was released and everything was thrown into tension. Things broke. A lot of things broke.

The rock beneath the crater, and for a thousand kilometres around, shattered, deeply and irrevocably, into huge blocks and arcs, metres and kilometres across. Those fault lines stayed in the basement of the rock forever – no matter that the land appeared to be healed by subsequent overlying deposits of material. The crust would remain vulnerable to movement along those lines. As well as later impacts reopening the old scars, the fractures were exploitable by other processes, such as volcanism and water.

Further below, in the plastic mantle where rock was hot and flowed on a timescale of millions of years, the cracks sealed up, but given the sudden removal of so much overburden, the mantle also pushed up into the absence left behind. The crater was initially even deeper than it appears now, but as the relaxing mantle pressed against the rigid crust above, all those broken lines and points of flexure shifted to accommodate the movement.

This had one further effect that would only be seen later. With the mantle now able to penetrate closer to the surface, retaining its heat but relieved of its pressure, it spontaneously melted. Then, by exploiting the shattered pathways to the surface, the less dense magma worked its way towards outlets around the crater rim. There are six separate sites in total where this happened, forming the earliest known volcanoes on Mars and the only major volcanic province in the southern highlands: Tyrrhena Patera and Hadriaca Patera in the north-east, and to the south and southwest Amphitrites Patera, Malea Patera, Peneus

Patera and Pityusa Patera. Whenever it was that these volcanoes began erupting, they continued to be active for hundreds of millions of years, so deep were the scars in the rock beneath them.

As the shockwave spread out from the impact, the energy was distributed across the wavefront. Surface waves – a marsquake – rattled through the existing landscape, bringing down crater walls, collapsing peaks and throwing them into valleys, losing energy all the while. The deep waves, travelling through the bedrock of the crust, shifted faults and moved blocks that measured kilometres across.

But Mars is a sphere. The shockwave radiating from Hellas reconvened on the exact opposite side of the planet. It's pure conjecture that this refocusing of impact energies had any effect, but Alba Mons – Mars's largest volcano by volume – is precisely there, 180 degrees from the Hellas ground zero. Hellas and Alba Mons are understood to be separated by at least a billion years of history, and their relative positions may be entirely coincidental. But there's a strong suspicion that Hellas produced a weakness that volcanism later exploited to create Alba Mons, having fractured the crust for tens of kilometres down.

The Hellas impact falls within a time period in the solar system known as the Late Heavy Bombardment. The evidence suggests that after the formation of the planets from coalescing planetary embryos, planetesimals and the other detritus of the debris disc, collision rates reduced from almost constant to sparse and sporadic. This happened not just for Mars but for all four planets in the inner solar system, as well as the Moon. Then, around four billion years ago, the rate of major impacts may have spiked again. This phenomenon was originally suggested to explain the formation of the dark lunar basins, which are craters that subsequently flooded with lava – none of the returned samples from those areas date older than 3.9 billion years.

In a simple model of planetary formation, cratering rates start high and quickly decrease, thereafter slowly dwindling away to currently observed rates. The model doesn't explain how a second peak in impacts could have occurred. We have, however, already suggested a mechanism that could explain it: the Grand Tack theory. The frozen outer reaches of the solar system still contain a great many objects, some of them very large indeed: Pluto, Eris, Haumea, Makemake, Sedna and Quaoar are all over 1,000 kilometres across. Add a myriad of other early solar system objects of all sizes, almost all rock and ice, and there was a ready store of projectiles to aim at targets lying sunward of them. Giant planets wandering outwards beyond the orbit of Mars would have undoubtedly perturbed the movements of any planetesimals that they encountered.

As Jupiter slid back out to its current orbit over the course of a few hundred thousand years, it would have rudely shoved Saturn towards the other gas giants, Uranus and Neptune. These planets in turn would have also pushed outwards, directly into the area occupied by the frozen accumulations of ice and dust that had previously escaped entrapment in the gravity wells of larger objects.

At this point, the orbits of these minor bodies would have changed, and many of them would have ended up heading on wildly elliptical paths towards the Sun. A barrage of unclaimed material, rich in volatiles, would have spun towards the centre of the solar system. Undoubtedly, Jupiter soaked up a great number of them, but many would have flashed past, accelerating all the way, until they turned around the Sun trailing comas of dust and gas, before heading back out to the cold depths of space beyond Neptune in their new, altered orbits.

The chances that any individual rock – we should probably call them comets, because that is what they were – would strike a planet were small. But given a couple of hundred million years of cycling from the very edge of the solar system to its heart and back again, those odds started to become more of a certainty. Impacts from these comets could quite possibly have added millions of tonnes of water and other volatile chemicals to the now-depleted Martian atmosphere.

There remains doubt as to whether the Late Heavy Bombardment ever took place: some of the lunar evidence has been questioned, although a sudden, discrete influx of large, water-rich meteorites does explain some of the physical and chemical anomalies we find on the rocky planets. It also makes a tidy conclusion to the Grand Tack theory. But it could also be that the impact rate did just slowly ebb away, and the creation of Hellas was merely an act of cosmic tidying up – one of the last great impacts of the early solar system.

THE START OF THE NOACHIAN

The picture we have of Mars after the first half billion years of its life is inevitably provisional. Reaching back so far in time, peeling back younger layers to reveal the ancient landscape, painstakingly reconstructing it from behind the fault lines, craters and volcanoes that came later is far from a straightforward task.

But the image we've built up so far looks something like this. In the south, there was a heavily cratered surface, sitting kilometres higher than the land to the north. There were the scars of relatively new craters – Argyre, Chryse and the great fist-mark of Hellas – plain and fresh, and of such depth that lava eventually broke through the cracks on the crater floor and puddled as sheets of infill. In addition, Hellas developed discrete volcanic sites on the crater rim, where explosive eruptions – highly likely given the reaction of hot lava meeting groundwater – created hot, thick ash clouds that rushed down from the summits, leaving in their wake dense, metres-thick, heat-welded jumbles of water-rich glass fragments and pulverised lava. Geysers and gases spewed from the ground – water, carbon dioxide and sulphides – and mixed with the atmosphere. The air pressure, outside of spikes caused by huge impacts, was somewhere in the region of one or two bars, and the temperature fluctuated wildly around a slowly cooling average.

There was an ocean, a mighty ocean that not only covered the northern half of the globe, but also turned the deep craters in the south into inland seas. And as difficult as it is to imagine that ocean and those seas as temporary, they were there at the whim of the local environmental conditions. They had disappeared before, repeatedly: one big meteorite strike and the soaring temperatures would evaporate every last trace of surface water and turn it into atmosphere. When it eventually came down again as rain, there would always be less of it than before, because some of it would have bled away into space.

Those times of great disruption were largely over, though smaller meteorites still fell, creating craters tens or hundreds of kilometres in diameter. Those that struck the land threw ejecta out in great blankets. Those that hit water caused tsunamis hundreds of metres high that washed across the shorelines and travelled far inland, up valleys and down again, carrying off all the sediment they could hold, which they dumped back into the oceans as they quietened.

Those ocean sediments formed stratified layers of rock: gently laid sandstones and mudstones made of eroded material from the highlands deposited in more placid times, interleaved with layers of churned flood deposits from the ebb and flow of impact-induced tsunamis. Meanwhile, on land, flood basalts – immense, region-wide eruptions of very fluid lava from fissures fed directly from the mantle–crust boundary – accumulated huge thicknesses of black rock. These were cut with lobe-fronted sheets of locally erupted lava from volcanoes, impact debris and wind-blown ash. The land was scoured from above, eroded by corrosive rain tainted with sulphur and carbonic acids.

In a sky that was primarily water vapour and carbon dioxide, there were bands and stacks of clouds, and possibly even a permanent fog that reached from ground level to the highest peaks. If the clouds did part, the revealed sky would be blue, because an atmosphere scatters shorter (blue) wavelengths of light more than longer (red) ones, effectively spreading that colour across the sky.

Rain-fed rivers cut through the surface deposits of broken ejecta and volcanic ash and used the debris they carried to grind away at the valley sides and riverbeds, carving channels and gorges in even the strongest rocks. They flowed downhill: away from the region dominated by the slope of the Great Dichotomy, water accumulated in craters or found its way into the faults that split the surface, where it was added to the deep reservoir of groundwater that was connected underground to the northern ocean.

From equator to north pole the ocean was a kilometre deep, with currents that carried the warm equatorial water to the pole, and transported the colder water back to the shore at the Dichotomy, making the equator more temperate than it might otherwise have been. And the weather: storms, immense and powerful, swept in off the northern ocean, bearing rain which turned to snow as the fronts moved inland and were pushed higher over the rising ground.

What about below the seabed? The basement rocks were brutally cracked, and water, seeking its lowest level, migrated into these cracks, filling up every available space, descending and heating up until it became too hot to stay liquid. Dissolved salts and other minerals filtered both up and down through the open matrix of loose sediments and volcanic debris, gluing them together to form solid rock. Hydrothermal vents formed along fault lines: boiling, mineral-laden water squirted into the cold depths of the sea. As we'll see shortly, if we ever want to find evidence of life on Mars, the remnants of the Noachian ocean is where we ought to look.

It is important to remember that catastrophic events – floods of lava and water, meteorites falling from the sky – were just punctuation to the main sentence. Hundreds or thousands of years might pass without incident. But it's just as important to realise that, given the long reach of the Noachian, those events were certain to take place. Within a span of a million years, a thousand once-in-a-thousand-years occurrences happened – a statistical certainty of instability and chaos contained within a semblance of what we might have considered normal.

The forces the planet had been subjected to already were monumental. And there was more still to come.

OBLIQUITY AND ECCENTRICITY

B efore we feel our way further into the Noachian, we need to talk about Mars's cavalier attitude towards both maintaining a stable orbit around the Sun and where its axis of rotation is pointing, because they have severe implications for the Martian climate at every point in its history.

Mars spins on its axis like a top; one revolution is a Martian day. It also circles the Sun once in a Martian year. So far, so normal. Because Mars's axis is tilted over, it has seasons, just like Earth does. When the north pole is angled towards the Sun, it's summer in the northern hemisphere; 334 sols later, or half a Martian year, the north pole is pointing away from the Sun and it's winter in the northern hemisphere.

Mars has two added complications to this entirely average planetary scenario. The first is this: no planet's orbit about the Sun is exactly circular. It is almost circular, but not quite, and this not-quiteness is called eccentricity. Comets, which have high eccentricity – that is, highly elliptical orbits – can be very distant from the Sun most of the time, well outside of Neptune's orbit, and then swing in fast and close to the Sun, inside the orbit of Mercury. Planetary orbits don't vary anywhere near as much, but Mars currently has the largest eccentricity of any planet barring Mercury, and its furthest and closest distances to the Sun vary by more than 40 million kilometres.

If summer coincides with the closest approach, it'll be unusually hot. Likewise, if winter comes around when Mars is at its furthest distance from the Sun, then it'll be colder than average. Moreover, the eccentricity itself changes gradually over time, driven by the gravity of other planets, chiefly by massive Jupiter. These periods of greater-to-lesser eccentricity occur in two cycles: a smaller one of 96,000 years and a much larger one of two million years.

The second complication is that the Great Dichotomy, however it was caused, altered the distribution of mass on Mars and left it unstable. As a result, the angle at which Mars tilts from straight up changes over time – and not by just a little, but by an enormous amount. The angle that a planet's rotational axis makes compared to the plane of its orbit is called its obliquity; this would be zero degrees if the axis was pointing straight up, and ninety degrees if the axis was lying flat. A planet is able to maintain a steady obliquity if it has a comparatively massive moon to stabilise it, but Mars doesn't have that: its two tiny moons do it no favours at all, so the axis wobbles back and forth.

Mars's current obliquity is twenty-five degrees, giving it moderately strong summer and winter seasons; it's been this way for a hundred thousand years, maybe more. But there have been times when the obliquity was much closer to zero, meaning that summer and winter were abolished for the duration. At other times the axial tilt has reached sixty degrees: under these conditions, the Sun would be higher in the sky at the poles than at the equator, and the polar regions would be subjected to 334 sols of freezing cold and dark, followed by 334 sols of burning heat and light. Year on year. Again, for hundreds of thousands of years.

The obliquity constantly changes, within a range, by a fraction of a degree a year. Over time, the changes accumulate, back and forth. A low-obliquity period, like Mars is in now, can last millions of years. But the swings between different obliquity ranges happen quickly, in a matter of a few thousand years. It's a chaotic system, feeding off of tiny changes in gravity due to Mars's eccentric orbit around the Sun. Going back even four million years – let alone four billion – to find out what the axial tilt would have been is a futile exercise.

While the effect of changing obliquity on the climate might not be predictable in terms of time, it is in terms of scale. Depending on the reinforcing factors of high obliquity and high eccentricity, the amount of energy from the Sun reaching Mars's poles can increase fourfold in summer. The opposite is also true. Mars has sustained million-year periods where low obliquities meant very little sunlight ever reached the poles: ice caps formed, locking up water that might have otherwise fallen as rain.

Young Mars was energetic enough, and had a thick enough atmosphere, to buffer its environment against the more extreme effects of its astronomical wanderings. But for older Mars – Mars of later, sparer periods – its entire climate came to depend on this chaotically changing obliquity.

INTRODUCING THARSIS

Tharsis is Mars's largest volcanic province, covering a quarter of the planet's entire surface. It has several descriptive names: Tharsis Rise, the Tharsis Bulge or the Tharsis Plateau. It's roughly 5,000 kilometres across, with a surface area of – depending on how we measure it – 30 million square kilometres. It straddles the Great Dichotomy and rises even higher than the southern highlands, up to 7 kilometres above datum: that's without considering the vast heights of the individual volcanoes that perch on it. The weight of Tharsis is estimated at a billion billion tonnes, and it's another feature, alongside the Dichotomy, that's large enough to interfere with Mars's axis of rotation. The weight of Tharsis has grown so great that the crust it rests on has failed in several places, opening up huge swarms of visible parallel cracks in the surface layers that presumably also extend into the depths.

Tharsis is a feature that dominates Mars. Our eye is inevitably drawn to this great bolus of rock that has erupted from the face of the planet, and the question of how it could possibly have got there and sustained itself over such an incredible length of time is one that will take some careful unpicking. There are several stories that need to be told here, but rather than two opposing tales, the formation of Tharsis is more of a rambling narrative that lasts far longer than it has any right to, complete with anecdotes and digressions.

Because as massive as it is, it has necessarily grown and changed over the four billion years of its existence. What we see of Tharsis today looks both permanent and ancient, but while some of it is very old, other parts are startlingly young. Centres of volcanic activity and tectonic uplift have shifted across the whole Tharsis region over time, either as a response to the vagaries in the precise direction of the mantle plume that is its primary cause, or as a result of the opening and closing of channels that allowed magma to rise to the surface as the crust flexed and filled. Tharsis is complicated. Different things have happened to different areas of it at the same time, and there's no straightforward presentation of the facts that fits.

We know that the only explanation for Tharsis is that it was, and continues to be, caused by a mantle plume: a single rising column of hot mantle material pushing against the base of the crust underneath it. We've already encountered mantle plumes in connection with the Great Dichotomy, but the plume associated with the Dichotomy was in the wrong place to power Tharsis. Either that plume moved from the south pole, or the mantle moved, or the crust moved, or all three – or another plume started. Is there any evidence of migration, either of a southern plume moving towards the equator, or of the crust having slid over it? Furthermore, why would it have stopped just the other side of the Great Dichotomy, rather than continuing on its journey?

To answer the first question: potentially, yes. There's a path from the south pole heading towards Tharsis, roughly along the longitude line eighty degrees west, that is suspiciously smooth given that this region is part of the heavily cratered southern highlands. There are large craters, visibly shallow and flat-bottomed – arguably filled – but the smaller ones we would have expected to see have mostly disappeared, giving a much younger age for the surface. Clearly, something buried the previous landscape, and close inspection reveals the telltale lobate edges of lava flows.

The first stage of Tharsis, then, was the migration of the plume northwards. Volcanism was limited to small, temporary structures expelling thick layers of water-rich ash and expansive fissure lavas, turned on and off as the plume passed by. Where the plume's progress slowed, the land rose upwards, both from the heat of expansion and from the deep-seated emplacements of magma within the crust itself adding to the bulk of the rock. Those intrusions pushed up through the fractured crust, forcing it apart and solidifying in place, time and time again, forcing the already-high highland to an even greater altitude. To either side of the plume's track, faults formed or were reactivated, and the movement cut across existing features.

But when the plume approached the Great Dichotomy boundary, it slowed its northward advance and then stalled. With some drift, both northward and southward, the plume has remained under Tharsis up to the present. It created the massive pile of rocks we see, and it is so large that it obscures almost all the evidence of its own earlier movement, much of its early history and also the boundary between northern and southern crusts. Tharsis's long rise might have started in the Noachian, but the process ran and ran, long after common sense would have expected it to stop.

The answer to our second question is that the plume stalled for a simple dynamic reason: a mantle convection current relies on the sinking of cool mantle material as much as it does on the rising of hot material. Where the crust was especially thick, in the southern hemisphere, it acted as insulation – the heat was sealed underneath and built up. Where it was thin, in the north, the heat radiated away more quickly through the rock, and cooled faster. As the plume crossed the Dichotomy, the mantle sank more quickly, and it trapped the rising plume just beyond the margin of the southern crust.

We'll inevitably return to this area, as it was – and remains – the most significant indicator of Mars's internal processes. It is the source of much of the planet's second atmosphere, the site of the largest rift valley and, due to its height, a place where equatorial glaciers form. Features unambiguously associated with water and ice occur side by side with those associated with fire and impact, and some of the landscapes found here are unique on Mars.

WE NEED TO TALK ABOUT WATER

However and wherever we look on Mars, we find evidence of water. With our satellites, we look for minerals that have been altered by water and turned into clays, and we find that they're abundant. With our rovers, we look for thin layers of exposed rock that could only have been transported by flowing water and deposited in lakes, and there they are. We see features on the surface that look so uncannily like dried-up riverbeds, feeding downhill into successively larger channels before washing out onto broad plains, that the only possible explanation for them is that they were indeed rivers. We see the results of catastrophic floods that have scoured the land into a chaotic jumble of broken boulders and carved shapes. We see shorelines where water once lapped.

But the water has gone. Where there were rivers, lakes, an ocean, all there is now is dust and ice. These are the bald facts: Mars was once a place where it was warm enough and pressurised enough to allow liquid water to flow at the surface, and now it's not. The Martian atmosphere of today is too thin and too cold to permit it. Under current conditions, water as ice sublimates directly into water as gas – it's not hot so we can hardly grace it with the label steam – and it precipitates out again from the vapour phase directly to the solid. We know this happens because we can see seasonal winter growth and summer retreat of the ice caps, north and south. Water and frozen carbon dioxide are transferred from one pole to the other on the ephemeral Martian winds.

And every year, in every summer–winter cycle, Mars loses a little more water as it moves from solid to gas. Vulnerable molecules, let loose in the atmosphere, are broken by the unfiltered sunlight, and the parts drift ever higher, eventually to be driven off into interplanetary space, scrubbed away from the tenuous top of the atmosphere by the constant scour of particles emanating from the Sun. For every year that passes, Mars gets drier and colder.

But in the Noachian, Mars had free water – enough that it would have been possible to sail around the world. The questions are, then, where did that water come from and where did it go?

As ever, we have only partial answers. We know some of the constraints. We know that Mars had to have had significant quantities of water: there are ancient shorelines that suggest times when large bodies of water existed, and the complex systems of rivers and lakes had to mean clouds and rain – a full water cycle of evaporation and precipitation. We know that when this cycle was working, the atmosphere had to be thick enough and warm enough to permit it. We can use our satellites to search the surface not just for minerals that still contain water bound up inside them, but also for ice, both on the surface and buried below it, shrouded by the rock and dust. We can, by crater counting, try to work out when the water flowed and when it stopped. Nothing is absolutely certain, but the story started something like this.

A planet, even one as short-changed on mass as Mars, only needs a tiny percentage of water within its original composition to be effectively rich with the stuff. These things are measured in metres of global equivalent layer: if you averaged the surface to remove all its high and low points, and spread out all the water across it, how deep would it be?

The exact figure is unknowable, but it is guessable. Taking an average from volatile-rich meteorites, perhaps half a per cent of the weight of Mars when it formed was water. If we gathered all of that together at the surface, we'd be kilometres deep in it. It's not the only version of the story of course; other interpretations present a much drier Mars, where the water layer was less than 100 metres deep, and Mars's atmosphere was always thin and cold. But a water-rich early Mars is the more likely scenario.

So where did the water go? When Mars melted, most of the volatiles – the water and the gases – boiled out to form the first atmosphere. Most, but not all. The mantle held on to some of its water, inside the minerals that make it up. We have supporting evidence for this: a wet mantle flows more easily and at lower temperatures than one which is devoid of water, allowing for mantle convection and the potential for plumes, such as those that may have formed Tharsis and the Great Dichotomy.

A lot of water was simply lost to space. We know that when a planet has no magnetic field the Sun's solar wind knocks out molecules from the top of the atmosphere. Over billions of years, an atmosphere can be stripped down to the soil.

Then there was the locking-in of water into minerals in the crust: clays are the obvious sink, but if conditions were right, carbonates are too. Clays are definitely present on much of the surface of Mars – although how thick these layers might be is yet to be determined – but there are also subsurface layers of clay minerals that would have formed through the action of underground water, and have been subsequently exposed at the floors and sides of large craters.

We can add to that all the other hydrated minerals that formed in cracks in the crust, deposited from hot water that percolated throughout the broken subsurface. Volcanic activity increases this, so wherever there was rising magma, there would inevitably have been mineral-rich water sluicing around nearby. Satellites have detected opal – hydrated silica – on the surface. There will be more to find beneath it.

On Mars, there was nothing – is nothing – to bind a loose surface together. Rain would immediately run off, dragging dust and sand downhill, the drops joining to form puddles, which then overflowed to form riverlets and then rivers, always seeking the lowest point. Craters would turn into lakes. Fissures in the ground would become sinkholes. The water might run to the sea, or it might disappear underground again, to be forced back to the surface in a different place, seeping through layers of rock as it went, altering minerals on the way.

So it travelled, round and round. And with each successive epoch there was a little less free water; a slow drying of Mars.

THE NORTHERN OCEAN

At a rough estimate, a third of Mars was covered by open water in the Noachian. Almost all of that water was in the northern hemisphere, in the lowlands beyond the Great Dichotomy. The ocean was vast and deep, and it sustained itself over hundreds of millions of years by a natural cycle of evaporation, condensation, precipitation and collection. In short, weather.

The atmosphere at this time had to be thick enough to support liquid water, otherwise there wouldn't have been so much of it for so long. Likewise, the temperature had to be locally high enough to prevent freezing, but not so high as to allow boiling. But before we dream of balmy days floating on the seas of Mars, we need to remember that the atmosphere was composed primarily of carbon dioxide and water vapour, with no free oxygen, and was potentially at a pressure of several bars.

As we're discovering, a warmer world is a wetter world. The storms on Mars would have been tremendous. Great torrents of rain would have descended on the bare ground, eroding the surface, seeping through cracks, running into rivers and ending up back in the sea. Most rain would have fallen within a narrow band to the south of the Dichotomy. Beyond that, in the shadow of the huge rise in elevation, there was very little rain – a desert. And further still, the south pole, and possibly ice.

When we look for evidence of this ocean and the rivers that fed it, of the ancient shorelines where the waves rolled up black beaches of basalt sand, of deltas of deposits laid down as the inflowing water dropped its sediment load, we find it. And more exotic tellings: the marks of impact-generated tsunamis, of broken boulders dragged up slopes by a wall of water and the wash as it receded, and later of stones that might have been carved from solid rock by glaciers, entrained in ice and dropped on the seabed as the icebergs melted. The story is surprising, but everything points to it being true.

When the ancient shoreline was first tentatively identified, it wasn't long before it was pointed out that portions of the sea level weren't at the same height. By the laws of nature and gravity, any large body of water finds its own level: add more water and it raises the whole ocean; take it away and sea level drops everywhere. A contemporaneous shoreline cannot be at two different heights.

But across four billion years? Mars has utterly changed during that time: the massive, ever-growing bulk of Tharsis flexed both the crust and the mantle beneath it, even affecting the position of Mars's axis of rotation. All this could account for the measured deviations in sea level. We're allowed our ocean.

The water that washed up on the beaches would have been of a different quality to our seas on Earth – no seaweed, no shells, no driftwood, no strand-line of debris. Twice a day there were weak solar tides, not significant lunar ones: Mars's two moons are mere peanuts and would have been unable to drag water from its basins and pile it up on its coasts.

There would have been storms, though. Waves are driven by the wind, and the wind is driven by differences in atmospheric pressure, which are ultimately caused by sunlight heating the surface more strongly when it's overhead than when it's at an oblique angle. The longer the wind blows on the same wave-front, the larger that wave will grow, and there were no shoals, islands or continents for the waves to break on in the northern ocean. The distance over which the wind can blow unobstructed, called the fetch, is essentially limitless in a circular sea. Our own southern ocean is feared by sailors for this reason. The Martian northern ocean would likewise have been a ferocious place.

And that's without the sporadic but inevitable meteorite impacts. Water, when hit hard enough, becomes as incompressible as rock. It forms the head of the hammer, even as it shatters, not cushioning the blow but transferring it into the seabed.

Imagining one of these events is difficult. All the energy of the meteorite would turn both the water above and rock below into superheated vapour that would want to do nothing else but expand. The shockwave would drive everything before it: the air, the ground and the sea. A circular wall of water would rear up, hundreds of metres high at the start, chasing and being chased by a blast of heat containing molten rock buoyed up on a raft of steam. The energy of the water wave would drop as it expanded out from the impact point. At twice its diameter, it would have less than a third of its original force, but moving so quickly that it would still strike the coast hard enough to break it. Pieces of rock would be torn out by their roots and carried inland. The coastline itself would change, with high ground laid low and riverbeds scoured deep by the surging water. The wave would bore inland until finally it could no longer support its uphill passage. The energy of it would dissipate, its momentum gone.

Then would come the great draining of the land as the water washed back into the sea, down those same rivers, pulling on those same rocks once again. Rip currents would tear at the sea-bed and gouge at river deposits laid down in less apocalyptic times. As the transient crater collapsed, the water would rush in, quenching the rock and violently turning into steam. There would be secondary waves, and the churning and roaring would continue. But for the most part, it would be over in a day. Until the next time.

The land would have been marked, though: the scars of those impact-created tsunami events were wild and chaotic, and only the long passage of time has softened them but not quite managed to erase them completely. The wave-carried rocks are still there, up on the mountainsides of Mars.

LIFE

We don't think we've found any Martian fossils. Back in 1996, there was a flurry of activity and some bold claims about the nature of the iron oxide particles discovered in one Martian meteorite, but these were – and remain – contentious.

ALH84001 – yes, it's that rock again, the one that haunts me – dates to around four billion years old, but it was only recently ejected from the surface of Mars, some seventeen million years ago, and even more recently entrained in a block of Antarctic ice, 13,000 years ago. This rock, when recovered from the ice and put under an electron microscope, revealed chains of tiny iron oxide crystals that would, in any terrestrial rock, shout loudly of a biological origin. Our oceans are full of magnetotactic bacteria: simple, single-celled organisms that navigate their position in the water column by sensing the direction and angle of the Earth's magnetic field using almost identical chains of crystals made of the iron oxide magnetite. Therefore, the investigating scientists thought the iron oxide chains in ALH84001 could be evidence for single-celled Martian life.

Sometimes the simplest answer is the right answer, and sometimes it isn't. If you try to emulate conditions on Mars four billion years ago, you can make iron crystals spontaneously line up in chains using purely chemical processes. Biology – life

– is not required. So we have precisely one Martian rock with tiny lines of iron in it and several explanations as to how they got there. The position is far from satisfactory and I can only apologise for that.

But the absence of evidence is not the evidence of absence. Just because we have not discovered, with our very limited explorations, anything that's obviously a fossil – a bone, a shell, a tooth – that in no way confirms that they're not there to be found. Fossil formation relies on several factors all occurring successfully and simultaneously, including burial rates, mineral content and mechanical and chemical events. Something that has lived and died is more likely to be preserved for millions, potentially billions of years if it's covered over quickly, pressed into a layer of rock and its soft parts subsequently mineralised. We don't know if the conditions for that ever existed on Mars.

Fossils can be hard to spot, too. Even the large ones. And especially if we don't exactly know what shape these novel early Martians might have taken. If they were individual single-celled bacteria, then there's not going to be anything to see until we get more samples back and put them under an electron microscope. But if, one day, a rover or an astronaut digs out a piece of rock – most likely from an old lake or seabed, or from the jumble of river deposits – and reveals something that is unmistakeably a fossil, even though we might not be able to say what it's a fossil of, ought we to be surprised?

Life is tough and tenuous, persistent and rare. In our world, it's common and yet precious. It grew from a series of poorly understood chemical reactions into a living, breathing skin that stretches almost pole to pole, from the highest peaks to the most

Hadean depths. Where it's most abundant, it's riotous. So where did it come from?

Back in 1952, Stanley Miller and Harold Urey conducted a simple but profound experiment. Using basic laboratory equipment, they made an assumption of early terrestrial conditions in terms of gases, added some water, heated it up and put a continuous electric spark across the mixture. They made all twenty amino acids necessary for life. They didn't make life itself, but life would need all those chemicals. It was a stunning result; it was also the first of many investigations into how complex organic chemicals might be formed – not just on Earth, but everywhere – and how that chemical soup might tip over into self-replicating organisms.

But we struggle with the whole notion of how to draw a line between self-sustaining chemical reactions and very simple life. We don't even know if there is a line, or whether putting down the markers on one side or other of it will mean we miss something important or end up mislabelling a purely physical process. The very simplest cells are nothing but a series of chemical reactions, with inputs and outputs. What could be the line is the way even the most basic life – stuff we might dismiss as slime – has the ability to self-replicate. Put simply, life creates copies of itself.

We have precisely one example to go on here: carbon-based amino acids, the basis of life on Earth. There's no guarantee this form of life is mirrored anywhere else, let alone on Mars. What appears to be important is the ability to chemically encode information in a way that's transferable, and the harnessing of some energy-producing reaction to power that transfer. We do that with DNA and aerobic respiration. But viruses don't have DNA and yet they manage to replicate and evolve: are they alive? Our arbitrary line says they are, but without living cells to infect and reproduce in, they're inert, so perhaps not.

We still don't know exactly what to look for, so what do we mean when we ask, 'Is there life on Mars?' In its baldest sense, we're asking if there's life that we'd recognise as life. Does it share the characteristics of life as we might understand it? Is it self-organising, self-replicating; does it conform to our idea of what being alive would be, even if it is just slime?

We've mapped Mars thoroughly from orbit and sent cameras to the surface. If aliens had done the same to us, their conclusion would be clear – yes, there's abundant life on Earth. But wherever we look on Mars, we don't see that abundance. We see instead a profound absence. There's nothing visible that we'd recognise as living. The conditions that exist on Mars now appear to be inimical to life as we understand it. But perhaps it exists within the ground, below the level where lethal radiation can penetrate. Perhaps it exists where there's free water – beneath the ice caps or deep under the permafrost. Until we look, we can't say, but we can test the hypothesis.

Extremophiles are organisms that can withstand conditions that would annihilate other forms of life. Some are adapted for high temperatures. Some for low temperatures, as they have internal fluids rich in natural antifreeze. Some for highly acidic or alkaline conditions. Some for crushing pressures. Some for radiation levels that ought to destroy DNA. Some live in rocks and breathe sulphur. Some live without oxygen. Some live without water.

Given that we can replicate the conditions on the surface of Mars in a laboratory, we can run experiments on our own extremophile organisms to see if they might survive there. The answer is yes: some of them can. Some of them can even survive being inside a simulated meteorite striking a surface at tens of kilometres per second.

These results are, however, disappointing. Given that we know that life might be able to exist on Mars, the fact that we cannot immediately detect it is suspicious. Where are the fields of lichen on the slopes of Olympus Mons? Where are the seasonal blooms of bacteria across the vast northern plains? Where are the odd, calcified structures of algae colonies jostling around the edges of the ice caps? Given that only the very simplest, hardiest life could exist, there would be little to no competition for the (admittedly scant) resources. If an organism can exploit an ecological niche, it'll usually exploit it to its maximum capacity, and we just don't see that on Mars. There's no obvious evidence of life at all.

We might have read it all wrong, though. Martian extremophiles could just be hanging on, millennium after millennium, in an environment too toxic even for them. Or they might never have evolved in the first place, or had evolved for different extremes and then vanished as Mars changed. Perhaps they were too slow to adapt to the ever-stricter climatic conditions. We know about mass extinctions: if Mars once had a tree of life, the changing environment might simply have taken an axe to it.

But what about conditions in the past? Might they have been more conducive to life? We find the very oldest examples of fossilised life on Earth in the preserved relics of a process that carries on today, deep down on the ocean bed. Hydrothermal vents are places where water percolates through cracks in the sea-floor and, having been heated by a subsurface source, re-emerges as a geyser of hot, mineral-laden water. These geysers are often so rich in dissolved minerals that the water is stained dark, which gives them their popular name, 'black smokers'.

The boiling water gushing out of the seabed into the cold, abyssal depths prompts the mineral load to precipitate out and form craggy columns of rock. Despite the darkness and the heat, isolated and complex ecosystems live and breed and fight and die on those steep slopes.

The exact same processes would have taken place on Noachian Mars. There was a deep ocean. There was water sinking down through a broken crust, heating up, dissolving minerals from the rock it travelled through and rising up again to be ejected forcibly into a cold, dark sea. If life is inevitable, then this is where we would have found it.

The question is: if raw chemicals made the leap to self-replication on Mars, how far along did the journey of life progress before the seas dried up? Mere freezing over of the surface would have mattered little to anything living hundreds or thousands of metres down. But it's as big a stretch from chemicals to self-replicating life as it is from single-celled organisms to creatures that might swim or crawl or float, let alone hunt, feel or think. Perhaps there was never anything more complicated than slime: when the seas finally sublimated away and the black smokers of Mars were deprived of the water that drove them, that was it. Global mass extinction, life irrevocably snuffed out.

Or perhaps life followed the heat and the water ever further underground, below the thick permafrost layer. Until we dig down, we'll struggle to know for sure. If we ever do find Martians, that's most likely where they will be.

There are reasons for our anticipation: we've found methane on Mars, when there's apparently no volcanic reason for it to be there. Methane can be a by-product of organic processes too, and the gas has a very short survival time in an atmosphere – ten to a hundred years – before it breaks down. Something on Mars is creating it. We don't know, yet, what it might be.

THARSIS RISES

As the Noachian progressed, more material was added to Tharsis, inside and out. The ground it started from was, on average, four kilometres above the northern plains, but the plume's presence, and the magma it created, pushed it higher still. The line of the Great Dichotomy became buried under Tharsis: not just the surface slope, but the structural depths too. The crust, fractured at a deep and fundamental level, shifted in response. If there was one aspect of Tharsis at this time that was as important as the volcanic eruptions, it was the creation – or reactivation – of crustal fractures.

Fault-bounded valleys, known as graben, are one of Mars's enduring features. They formed – just like they do on Earth – when kilometre-sized blocks rose or sank in response to tension within the crust. To make a graben, a central block sinks relative to the blocks either side of it, creating a wide, flat-bottomed valley with sharply sloping sides. Most grabens are suspiciously straight, as if designed by some intelligence: imagine bricks resting on a slowly inflating air bed, shifting and jostling together. Noachian Tharsis was both covered and surrounded by these valleys: sometimes complex, sometimes buried, the ground stretched as it gradually pushed upwards. Some parts rose much higher than others and created the ring of ridges around the high plateau of Syria Planum. These ridges remain today: the most significant is the Claritas ridge in the west, and only the north-east is unguarded, as the land slopes down towards the Chryse impact basin on the margin of the Great Dichotomy.

To the south-east of Tharsis lies the huge and deep Argyre crater – a lake at that time, just a little short of four billion years ago. The mid-to-late Noachian Argyre was fed by meltwater from the south pole, which ran through rivers that carved their indelible way through the highlands and ended up collecting in Argyre's large inland sea. The water appeared periodically to escape the almost 2,000 kilometre-wide basin, making its way north through the Uzboi Vallis and its nested chain of rivers – the Uzboi, the Ladon and the Morava. The Uzboi Vallis potentially started as a graben, but was modified later by the water flowing down through it. The river system itself was augmented by run-off from the burgeoning Tharsis uplands on its way out to the ocean: water always seeks the lowest point.

And it wasn't just water that poured off the Tharsis region. Lava emerging from fissures in the ground flooded out, covering the floors of the plateaus and filling craters until they were all but buried. Eruptions also belched steam and other gases – carbon dioxide and sulphur – into the air, changing the quality as well as the quantity of the atmosphere. The primordial mix that blanketed Mars straight after it melted had gone. What was there now was on its way to becoming a more nuanced, sparser secondary atmosphere in which the volcanic component was increasingly important for boosting the pressure. Since the core-generated magnetic field had failed, the solar wind had been stripping the atmosphere faster than it was being created below.

It's difficult to say whether there were any true volcanoes present on Tharsis in the beginning. Eruptions from a single, sustained vent can form mountains from a flat plain in a remarkably short time frame, but these seemed to have occurred later in Tharsis's long life. If there were examples back then, they've been either buried by subsequent lavas or levelled by ground movements. Instead, cracks in the ground spewed out copious floods of lava that flowed almost as freely as the water. The land bucked and rose, cracked and heaved, as far below it a planetary-scale fist of rising mantle pushed up against the underside of the crust.

LAKE ERIDANIA

One of the extraordinary characteristics of Mars is the sheer persistence of its features. When we point to a geological feature on Earth and call it old, it will almost certainly be less than a billion years old. Almost all of the rocks we see are far, far younger than that, even though they're still tens or hundreds of millions of years old. On Mars, the landscape is a mosaic of the genuinely ancient – billions of years old – with younger deposits on top.

So identifying a series of basins in the southern highlands as the site of a late Noachian inland lake, larger than any lake elsewhere in the solar system, is remarkable, but by Martian standards it is entirely reasonable.

Evidence for the existence of a lake in Eridania is threefold: firstly, there's its geography. Eridania is a series of ancient circular impact craters nested together to form a linked chain of low points, surrounded by the higher ground of two plateaus, or terrae: Cimmeria and Sirenum. The outflow from the lakes is where a huge river – the Ma'adim Vallis – once met the old crater rim. This low spillover point is where water would have crested the lake's edge and poured downhill in a torrent, towards the sea. The Ma'adim Vallis traced its way a thousand kilometres to the ocean, where the river spilled out onto the northern plains through the Gusev crater – the site of the Spirit rover landing in 2004.

The second piece of evidence is Eridania's relationship with the still-visible river channels that extend across the highland surface, in a band that starts at the Great Dichotomy and stretches some thirty degrees further south. Rivers were abundant in this region, but we don't see them in the area covered by Eridania: the river valleys finish at a much higher level, almost as if they flowed into a large body of existing water.

The third sign is the presence of hundreds of metres of clay deposits within the craters, which can be best interpreted as having been laid down over a significant period of time – hundreds of millions of years – in a deep, replenishing body of water. These deposits vary in thickness across the lake bed, and where they've been excavated by more recent craters, the clay appears to go as deep as two kilometres.

The clays came from erosion of the highlands. Water mechanically and chemically broke down the rock of the volcanic hinterland and carried it to the cold, still depths of the lake. Rivers reduced the particles' size as they flowed downstream – boulders, cobbles and pebbles clashed in the energetic headwaters to form the sands, silts and muds found in mature lowland streams. Once the sediment found its way into the lake, it settled. If there had been filter-feeders present, they would have collectively siphoned their way through vast quantities of water in the hopes of a meal. The regurgitated clay-sized particles would have ended up as pellets, stuck together with mucus. This would have made the settling process much quicker and more effective, but in the absence of sponges, shellfish, worms and other crustacean-like creatures, the clays had to settle naturally, through the water column, onto the lake bed.

Standing on the slowly curving shoreline of Eridania, we might have watched the rain clouds roll in from the north. The far side of the lake could easily be one, two, three hundred kilometres away, but we wouldn't be able to see it, as anything further than a few kilometres would be below the foreshortened Martian horizon. Waves would climb to improbable heights in the low gravity before looping over and breaking on the headlands and in the bays. The land around us would be necessarily barren and scarred – nothing but layer after layer of dark volcanic rock and impact ejecta, piled up into mountains and eroded back down into battered, worn highlands and filled-in hollows. All around the lake, silver rivers would run down across shallow beaches of black grit and into the water.

The rivers were not the sum total of water entering Eridania, though. Water was percolating through the basement cracks at the bottom of the craters and feeding the lake through underground springs that might have been superior in volume to the surface water. Water also left the lake in the same way, heading underground and into the crust.

However, what we know of Mars is that where there are deep impact basins, there's often volcanic activity. The twin effects of decreased pressure on the mantle below and increased infiltration of water often meant that rock spontaneously melted at depth and rose up to resurface the floor of the crater, seeping out along the same cracks made by the impact.

Now that the water is long gone and the lake bed is exposed to view, this is what we find: that hydrothermal activity – those black smoker vents pouring out mineral-rich hot water – was present. Not only that, but even though Eridania was flooded to depths of a kilometre, lava was forcing its way out into the lake-bed sediments. That both water and lava can coexist is perhaps counter-intuitive, but once the lava forms a supercooled crust on its outside, it can remain molten inside and still flow. Likewise, a skin of steam on the water side stops the whole lake from quenching the lava. In several places on Eridania's floor, we're treated to the sight of once-wet lake-bed clays cut into a chaotic terrain of islands by dense submarine lava flows.

The longevity of the Martian landscape means that it can be difficult to unravel the sequence of events that took place both above and below the surface of the lake. If the water retreated during dry spells, then the lake as a whole would have grown much saltier, and evaporites – water-soluble minerals such as salts, carbonates and gypsum – would have formed at the shore-line. But the craters would have been flooded again when the climate turned. Eruptions increased the temperature of the lake as a whole, but given the latitude and height of the area, ice would also have covered some or all of the surface at various times.

Extreme weather events meant that water periodically over-topped the basin spillway, sending a flood down the Ma'adim Vallis and turning what was a steady if unimpressive river into a torrent that scoured the channel sides and floor and pushed debris down to the vallis's end, the Gusev crater, and across the northern plains.

That this huge lake ended up a dried husk, slowly and relent-lessly drained by a changing climate and a thinning atmosphere, was inevitable. As the shoreline contracted, the clays shrank and subsequent lavas laid down ash and solid flows across the saturated ground. Meteorites smacked into the lake bed and splashed out sloppy ejecta as mudflows, exposing the layers of deposited and sometimes altered sediment in their craters. Dust eventually covered everything and frost gnawed at the exposed rock.

But that there was once such a place is enough. It represents a snapshot of what conditions were like across large areas of Mars as the Noachian slid towards the Hesperian: wet, broken, active, yet always with the spectre of a cold, still, airless future ahead of it.

PART FIVE

THE HESPERIAN

(3.7–3 billion years
before the present)

THE HESPERIAN CLIMATE CHANGE

The beginning of the Hesperian era is defined by a dramatic change in Mars's climate. The late Noachian had been mild and generous – rich with water and covered by a thick blanket of atmosphere. It would have been still, obviously, uninhabitable by us, but it had potential.

Sustained volcanism changed everything. Once magma broke through to the surface in sufficient quantities, gases poured into the atmosphere and the quality of the air altered. One single volcanic eruption might have had a significant local effect on air quality – in terms of not only dust and ash, water and carbon dioxide, but also sulphur dioxide and other sulphur compounds, hydrogen fluoride, hydrogen chloride and carbon monoxide. But more eruptions, either all at once or intermittently and repeatedly over tens of millions of years, fundamentally altered the global balance.

The ocean, up until this point, was mildly alkaline, and conditions were conducive to the production and laying down of clays. While extra carbon dioxide and water vapour bolstered the thinning Martian atmosphere, the other volcanic gases readily dissolved in rain to form acid solutions. The slow drip-drip of acid then fell on the highlands and found its way into the northern ocean and the crater-lakes. It shifted the balance from clay to the laying down of sulphates – gypsum and other salts.

Mars's water gradually became more acidic, and it stayed

that way from then on. Whether or not that ended any nascent Martian life is unknown, but extremophiles are usually highly specialised organisms. Those that are resistant to heat are not usually also resistant to acid, and vice versa.

It's thought that Hesperian-age volcanic rock eventually covered almost a third of Mars – much of it in the form of flood basalts in the northern lowlands, as well as on Tharsis and on the floors of highland craters, notably Hellas. Much of this seems to have started in the late Noachian and intensified in the early Hesperian.

On its own, volcanism would have raised the temperature of Mars a little, but there was a greater and more dramatic change happening far above sea level. The volcanic sulphur dioxide aerosols high in the atmosphere reflected sunlight back out into space, causing the air below, and therefore the ground, to cool rapidly. The gases tended to wash out quickly – within decades – and turn into acid rain, but because the volcanoes constantly belched out more sulphur, there was always enough to keep the levels up. The immediate effect was a global chilling. Frost formed where it had never formed before. High-latitude lakes iced over. The longer those conditions persisted, the deeper the cold bit. The first frosts on the equator signalled that the climate had tipped in favour of winter.

Water-saturated ground turned from being merely ice-covered into permafrost. Ice penetrated tens of metres deep. Lakes froze from top to bottom. Ice caps formed, grew larger and crept towards the lower latitudes. In the highlands, snow fell, accumulated and didn't melt in summer. Persistent, year-round ice turned into glaciers.

Locking away water as ice interrupted the cycle of evaporation–condensation–precipitation. The northern ocean began to contract. The seabed was exposed. Lakes dried up. Most critically, the water vapour that kept the air pressure high was now trapped in crystalline ice, and the thinning atmosphere became even worse at holding heat.

The effect was cumulative and reinforcing. Throughout the Hesperian, Mars turned from warm, wet and pressurised to cool, dry and gasping. Each puff of yellow volcanic gas caused another fractional drop in temperature, which created more ice crystals, which led to a drop in air pressure, which meant the heat capacity of the atmosphere fell a little further. The colder temperatures meant less evaporation, which meant less rain to wash out the sulphur aerosols, which then persisted for longer and led to further cooling.

If Martian volcanism had been for but a season, the climate might well have recovered. But after the initial surge of eruptions had altered both sky and land, substantial periodic volcanism throughout the Hesperian and beyond locked in those changes. There were temporary, million-year-long revivals: as the planet's obliquity turned to greater angles and the ice caps retreated, the planet warmed and water flowed again. Meteorites periodically struck the old ocean floor and melted enough of the ice and generated enough heat for lakes and even seas to form. But this was always in the context of the overall trend. It was never enough to reset the clock. The slowing of the astronomical bombardment played its part, as did the solar wind gradually stripping away the atmosphere. There was no turning back.

Mars was entering a new phase, one dominated by fire and ice.

THE BEGINNING OF
THE CRYOSPHERE

That ice can be as all-pervasive as air and water is a notion that those who've experienced life at the poles will find easier to grasp than the rest of us, but the consequences for a previously warm planet that freezes from the surface downwards are profound and long-lasting. The cold spell that started in the Hesperian has lasted, on and off, for over three billion years and into the present day, so we need to understand the effect it has had in order to continue with Mars's story.

As the cold penetrated water-soaked surface sediments, it froze the ground solid. Water, when frozen, occupies almost 10 per cent more space than when it's liquid. This expansion caused Mars's surface to buckle and heave into new shapes, disrupting the river- and lake-formed features and blurring them. Periodic warming collapsed the permafrost, causing slumping, landslides, surface sinkholes and temporary rivers. As colder conditions returned, everything slowly froze again. Landscapes changed, their previous forms erased.

Volcanic rock was also affected, although differently. When hot rock cools, it contracts, breaking along lines of stress. Joints form through the layers and water can seep down into them. Newly formed Martian ice forced those joints wider and, at the surface, shattered the rock into sharp shards. Below, where the tension had nowhere to go, the pressure inside the rock grew enormously.

The permafrost bit hard and deep. As the years turned into millennia, the ice layer penetrated further down into the crust, creating a thick – and watertight – cap over much of Mars, potentially a kilometre or more deep. This cryosphere of ice replaced much of the hydrosphere of water – although not completely, as deep down in the warm darkness, underground currents still flowed. But above, the water cycle of evaporation–condensation–precipitation became locked up in an ice cycle of sublimation and deposition. The ice did flow, but only slowly, and it needed to be renewed by fresh falls of snow. When it wasn't, it was scoured from the top by the wind and carried away, leaving only bare frozen ground behind.

Exposed free water at the surface became ephemeral. The northern ocean armoured itself with ice. The water underneath remained protected for a while, but as the biting cold continued, liquid water froze onto the underside of the floating ice, until it had all solidified. The water currents that might have driven the ice above to break up and to flow, ceased. Deeper craters held on to their water for longer, but eventually they too succumbed.

There were limits to the march of the ice. The permafrost extended down only so far as the freezing cold won out over the rising heat of the still-warm mantle. In active volcanic areas, where the heat-flow was much higher, even the surface might have remained unfrozen, for a while at least. Hydrothermal vents stuttered on, but as they were reliant on the percolation of water down to the warm rock far below, they dwindled to hot springs and geysers, and then to nothing. The life that might have been associated with them? Unless it retreated far underground, it would have been extinguished.

Over the ocean, meteorite impacts, if large enough, punched their way through the ice and into the ground below, generating tsunamis of ice, water, steam and molten rock. On land, an impact in the permafrost unleashed an ejecta blanket of fluid, muddy debris that then froze in place. The crater stayed warm as the heat dissipated and became an oasis in a frozen desert for years, perhaps centuries – or even longer if the crater was later inundated with lava from below – but in the end this would have frozen over too.

Volcanoes could also erupt below ice sheets, either as a new vent, or in the case of an existing volcano, beneath a glacier that occupied its summit. Heating the base of the glacier would have created huge quantities of meltwater which, once it saturated the ground underneath, had nowhere to go except out, any way it could. Those spontaneous and abrupt flooding events scoured existing channels down to the bedrock and below, and created new ones where none existed before. The water may have been fleeting, but the scarring effect was permanent.

There was a third way in which liquid water could have appeared at the surface of frozen Mars: for this we will have to go deep underground.

VALLES AND CHAOSES

There are thirty named chaos terrains on Mars, each characterised by the same jumble of flat-topped blocks, kilometres wide, tilted at every angle and separated by steep-cut valleys hundreds of metres deep. They are each set into a depression and associated with a vallis – a broad river valley with obvious flowing-water features that seem to empty away from the chaos. Some of the chaoses are huge, hundreds of kilometres across in total. They look like unhealed scars on the landscape: crazed and creased and thin.

There are almost as many theories of chaos formation as there are chaoses, but they all seem to coalesce around the sudden and repeated upwelling of great volumes of groundwater, directly out of the subsurface, that broke through the thick lid of permafrost. Such a phenomenon isn't really found anywhere else, and certainly not on Earth, but the chaos landforms are reminiscent of places where we know sudden flooding has happened in the past, so what follows is very much a best guess. It's dramatic and slightly complicated, but it does make a good story.

As Mars froze solid above, enough heat rose from the mantle for the crust to remain warm and thus for water inside it to stay liquid. That water may have been trapped under a solid, waterproof layer of ice and rock, but it still flowed through the cracks and pores of the fractured subsurface, rising and falling, swirling in underground currents, driven by differences in temperature and pressure. Water was potentially added to the system at the poles through the melting of ice beneath the ice caps, but all the circulation took place at depths of several kilometres. Those vast movements seemed destined to stay hidden; if they climbed up far enough to touch the permafrost layer, their flows became sluggish and icy.

Yet they were under enormous pressure: the water was trapped and squeezed by the weight of the frozen rock above. If we could have gone there and drilled into it, we would have been rewarded with a gusher of warm, salty water rising high into the thin Martian sky, freezing in the sparse, sub-zero air, falling as snow and simultaneously sublimating into vapour as it fell. Whatever the initiation event was, be it volcanic, tectonic, meteoritic or simply hydrostatic (the rock failing due to the fantastic strain the water placed on it), once some of the water had managed to punch through the ice layer, the breach inevitably widened and let out more. What this would have looked like seems almost impossible to describe, yet we have to try.

At first, there would be a rumble, a marsquake, and then a jet of water as hard as a steel rod would cut its way through the ground and roar out. Rock, eroded from the walls of the initial shaft and entrained in the jet, would widen the channel quickly and weaken the land around it. One failure would grow into multiple points of breakthrough – fractures spidering out deep underground and breaching the surface. The noise and the vibration would grow and grow until, suddenly, whole mountains would move. The seemingly solid ground would quiver; it would all sink, but some parts would sink more than others. Slabs would break off, tilt and tip, and water would be everywhere. A vast bolus of water – a sea's worth, a spontaneous flood – would leap into existence where moments before there had been nothing but bare, dry, frozen rock. It would boil out of the ground, an explosion of liquid.

That seething, churning, mud- and rock-laden warm-water wave would gouge its way to the lowlands, digging out a channel, carving its sides, shaping its passage and only slowly losing speed and potency as it travelled. Eventually, tens or hundreds of kilometres downriver, it would subside. As the wave of sediment-laden water poured into the craters and out over the plains, the heavy material would drop out first, and the dust last, to form a single graded deposit running from coarse at the bottom to fine at the top.

Behind it, at its source, the first burst of groundwater would slacken, and as the throughput slowed, the ground would start to refreeze. The permafrosted rock layer, a huge reservoir of cold, would rob the water of its energy and the flood would turn into a river, a spring and then shut off like a tap.

The surface water, still seeking the lowest point, would pond in craters and channels, cooling and resting. Ice would form above and below until the water froze solid. The mud and the rock would freeze with it, potentially forming rock glaciers – huge, ice-cored rock flows where the ice is protected from sublimation or melting by a covering of debris. Weeks and months would pass, and although the landscape would be utterly changed, all of the surface water would have already either frozen in place or slipped back underground. The vents that brought it to the surface would be stopped with ice. The ground would quieten.

Beneath the permafrost layer, the underground water would begin to refill the depleted joints and pores, like a desiccated sponge swelling back up. With the weaknesses already created and ready to exploit in the rock above, the critical threshold of pressure to breach the surface would be lowered. The chaos may have to wait a month, a year, a decade or a century before the ground cracked open again, but it would, allowing more water to spew out along the already-formed channels, to wash them clear and widen them further. And so on, until the wound of the chaos closed over completely and the aquifer below was left so empty that it could never refill enough to break through again.

The chaos landscape that was left would be sharp and hard, as yet uneroded or smoothed over by dust. City-sized slabs of the earlier surface would lie where they'd fallen, heaved up and over by the inconstant ground. Separating them would be scoured land and flood-washed canyons. There would be new valles that marked the passage of the flood water, which had gone where it wanted, exploiting earlier lines of weakness and making new ones. Out on the plains, new layers of water-rich flood sediments would have settled and then frozen. Inside craters, the infill would have raised the crater floor, burying features like rings and central peaks, potentially leaving only the very rims exposed.

The time of chaoses eventually ceased, but they rumbled on through the Hesperian. The outwash from them coated the northern plains all the way up to the north pole. The Vastitas Borealis, the Acidalia Planitia, the Utopia Planitia and the Arcadia Planitia are home to the majority of the exposed Hesperian-age sediments, along with the Chryse Planitia, which lies at the mouth of the Valles Marineris. They are anywhere from tens to thousands of metres thick – an extraordinary amount of material transferred from the highlands to the lowlands.

Time eventually softened the chaoses. Frozen debris slumped down, and fresh frost attacked the cliffs and turned them into curtains of scree. The valles blurred with wind-blown loose sediment: dunes formed on the riverbeds and sand piled up on the leeside of obstacles.

The floodplains now crackle with frost: they heave and sag in season with the surface temperatures, and add some small measure of relief to an otherwise featureless horizon.

HERE ARE GIANTS

There are twenty-eight large volcanoes or clusters of volcanoes on Mars, and many more smaller vents and cones. Volcanoes on Mars run the gamut of sizes, from the biggest beasts in the solar system to structures that aren't even bumps in the ground. Most of the very largest are contained within Tharsis. Purists will tell you that next-door Olympus Mons, which counts as the absolute tallest volcano and the second tallest mountain in the solar system, isn't part of Tharsis. But whatever process built Tharsis and the region's mighty volcanoes – Alba Mons, Arsia Mons, Ascraeus Mons, Pavonis Mons, Uranius Mons, Uranius Tholus, Ceraunius Tholus, Biblis Tholus, Ulysses Tholus, Tharsis Tholus – also sowed Olympus Mons from the same seed.

All volcanoes start the same way. Magma forces its way up through the ground from a pressurised reservoir somewhere deep underneath. When it breaks through to the surface, volcanic material – whether it's lava or more explosive products like fine ash or cinder bombs – is expelled from the vent. When more of this collects closer to the vent than further away from it, a mound forms. Continuing or repeated eruptions build on this unpromising beginning, layer after layer, metre after metre. That's it – year on year, century on century, millennium on millennium, the vent widens as the rock of its throat is scoured by the upward force of magma, and it rises up as the volcanic material accumulates around it. This process continues for as long as the volcano is fed fresh magma from below.

Small volcanoes can grow quickly and fall extinct just as rapidly. The more substantial ones can go through cycles of waking and sleeping, with periods of dormancy that are sometimes so long that erosion starts to wear away the summit. The magma chamber below the volcano might retreat, cool, contract and solidify; any empty space left inside the mountain will eventually be filled from above as the summit of the volcano collapses into it, erasing the volcanic crater and leaving the cliff-bounded, flat-bottomed structure we call a caldera.

Some volcanoes have just one caldera. Others have evidence of several, and crater counting on their floors shows that these volcanoes have lived and died and lived again through hundreds of millions of years, sometimes longer. Such longevity and such resurrections are only possible because of Mars's unique set of circumstances. On Earth, a more mobile crust shifts the volcano – and the magma chamber that supplies it – away from its mantle heating source. On Mars, the volcano might remain over the same hot spot forever.

The oldest volcanoes appear to be those in the southern highlands; at least, they stopped erupting first and are left as scabs surrounded by heavily cratered lava fields. Worn down by billions of years of erosion by rain, flowing water, ice, dust-laden wind and periodic meteorite strikes, they're barely recognisable as volcanoes at all. The layers of lava that surround them, and their rimless collapsed calderas, are all that remain.

Of the younger volcanoes, here's the slightly disappointing truth: if we were to visit one, we'd struggle to realise that we were on the flanks of a volcano at all, let alone one with a kilometres-high peak. From above, they look hugely impressive: great, broad-shouldered structures with caldera-punched summits. Arsia Mons boasts a 100-kilometre-diameter caldera that lies a full kilometre below the highest point of the rim above it. Lava flows fan out on all sides of the lower slopes in scallop-edged formations, stacked up one on the other, raising the volcano five, ten, fifteen kilometres above the surrounding plain. Other features – side vents, lava tubes, collapse structures like landslides and chain pits – relieve the monotony of one step of solid lava after another.

But because lava on Mars seems to have been mostly very fluid, the volcanoes there are nearly flat. The average slope on Ascraeus Mons is four degrees. For every 100 metres walked, we'd climb 6.5 metres. It's not nothing, but neither is it the steep-sided cone of popular imagination.

Standing on the lowest slopes of Ascraeus, looking up in the direction of the summit, all we'd see is the rise, not the top. And standing at the top – 18 kilometres above datum – we'd have no view at all. We could look down into the caldera, but the rest of the compass circle would be a near-level horizon a few kilometres distant. And even if we knew the direction of the next volcano in line, Pavonis, which is itself 14 kilometres tall, the curve of the planet would prevent us from seeing it at all – either the base or the summit. From our high vantage point, the horizon would be just over 350 kilometres away, and Pavonis is 800 kilometres distant. As mighty as the volcanoes of Tharsis are, they're more or less invisible from the ground. The climbs are gentle, the summits hidden and the highest points marked by sudden kilometre-tall cliffs falling away beneath us.

Tharsis doesn't appear to have had any volcanoes as such until the Hesperian; before then, all the gain in height and weight was due to uplift of the crust, subsurface intrusions and eruptions from rifts in the ground. Single-point volcanism was new. When it did happen, it started small, with Biblis Tholus and Ulysses Tholus in the east, Tharsis Tholus in the west and Ceraunius Tholus and Uranius Mons in the north. These are all significant volcanic structures, but they're utterly dwarfed by the three west-of-centre Tharsis Montes volcanoes, Pavonis, Ascraeus and Arsia Mons, by Alba Mons in the north and especially by Olympus Mons in the far west, all of which grew later.

The volcanoes of Tharsis are strange by nature: complex and ancient and so very persistent. Given the Methuselah-like lives of these vast blemishes on Mars's skin, to call them all extinct is an act of hubris. For now, they're apparently quiet, and that's all.

VALLES MARINERIS

Valles Marineris is another of those ludicrously oversized features that Mars seems to specialise in. It is a series of connected, steep-sided, flat-bottomed trenches that stretch for 2,000 kilometres from east to west through the almost-but-not-quite middle of the Tharsis region. These trenches – one of which is properly called a chasma, the plural being chasmata – are 200 kilometres wide, and their floors can lie 10 kilometres below their rims.

Of course this feature has to do with Tharsis – the sheer weight and size of it – but there's more to Valles Marineris than that: there are some spectacular deep-crustal elements that combine to make this vast crack in the Martian surface simultaneously inevitable and awe-inspiring. It even created a lake system that led to some of the most significant flooding events Mars would ever see.

Before we consider how Valles Marineris formed, we ought to look at a map of it, better to contemplate its heights, depths and widths. From north to south, there are four lines of chasmata. The first consists of the shortest chasmata: Echus Chasma and Hebes Chasma, two just-unconnected valleys in the north and off to the west. Then there are three parallel scars, each with a breach in their wall into the next: Ophir Chasma, Candor Chasma and then the largest and longest system, which is really one valley but is divided up into three sections: Coprates Chasma in the east, Ius Chasma in the west and Melas Chasma in the centre.

At the far western end is the uniquely Martian landscape of the Noctis Labyrinthus, which looks like a plate that's been dropped onto a stone floor and then clumsily glued together again. It's a shattered region composed of cliffs with slumped and broken sides that tumble into valleys, separating high islands of the original surface. These smaller chasmata branch and split and rejoin, gradually shallowing to the far west, as they head towards the high ridge of the Claritas Rise.

At the far eastern end of Coprates Chasma there are three additional trenches, but these are aligned differently and they all meet at their eastern extremities: Ganges Chasma in the north and Capri and Eos Chasmata in the south open into the Aurorae Chaos.

The sides of an individual chasma are almost vertical – in the case of Candor, they descend 7 kilometres in one single step. The sheer cliffs have been subject to later collapse – blank walls transformed by footings of frost-shattered debris, the broken rock smoothed over with washed-down highland sediment and further blurred by wind-blown dust. But nothing can fully hide these extraordinary wounds. They are too big, too obvious and they crave an explanation.

Again, like so many phenomena on Mars, there are several competing explanations. Perhaps the chasmata are rift valleys, formed when the crust to the north and the south moved apart and stretched the ground in between – ground that might have been superficially healed by overlying Tharsis lavas, but below was still all fractured crust, kilometres-wide blocks of rock broken off from their neighbours by huge impacts in Mars's early history.

Or maybe the chasmata were formed by the action of flood basalts scouring channels in Mars's crust as if they were rivers – hot rock pouring from west to east, tearing at both the walls and the bed, carrying and melting the debris as it passed.

Alternatively, they could be the result of explosive outbursts of groundwater, activated by heat rising from below, that undermined the surface and dragged all the overlying rock and dust away in a series of cataclysmic, biblical floods that spent themselves in the chaotic terrain of western Chryse.

For certain, there are volcanic and river deposits within Valles Marineris. There are small cindery cones on the floors of some of the chasmata, and there's evidence of lava flows. Water has clearly passed down the valley system, too. There's more than enough evidence for both of these hypotheses.

None of these stories quite satisfy, though. Why there? Why then? There are compelling reasons to believe that the chasmata of Tharsis were the result of deep-crustal movements that related to two utterly uncoincidental factors: the steadily growing size of Tharsis and the position of the Great Dichotomy.

We know that Tharsis formed on top of a stationary mantle plume, which added material above due to volcanic lavas and below due to intrusive magma. Tharsis rose and stretched and flexed throughout the Noachian and into the Hesperian, with centres of activity that slowly switched from south to north, to east to west, and back, but that always accumulated weight, always grew and spread.

The great new mass didn't sit lightly on the ancient crust. It bore down on it, causing it to sink deeper into the mantle. But it didn't sink evenly, because the weight of Tharsis straddled the Great Dichotomy. The crust to the south is an average of 60 kilometres thick; the crust to the north is half that. They sank at different rates in response to the same weight, creating a line of violent tension that radiated all the way up from the bottom of the crust, where it joined the mantle, to the very top, where Tharsis was unthinkingly adding more lava and jamming in more intrusive rocks.

The crust, and Tharsis above it, cracked in a series of parallel tears. Great rectangular blocks, aligned with the surface cracks, fell into the zone of maximum tension below. Lubricated by magma, they sank down an unprecedented distance and found a new, stable floor, kilometres below.

Whatever the trigger for the collapse of the crust beneath Tharsis, the Valles Marineris formed quickly. The early Hesperian seems the most likely time for the initial stages of collapse, given the ages of the lavas at the top of the formations.

But there's more. The measured depths of the chasmata appear to be much greater than required to balance the weight of the sinking blocks, even for something as huge as Tharsis. They should have only shifted downwards a couple of kilometres at most, so why do we find them much lower than this? This time, the answer lies not below but above ground.

The newly formed chasmata obviously provided low points on the surface of Mars. We know that water was present at the time: we can still see evidence of river systems that fed into the chasmata from the high Tharsis plains, but groundwater was more likely to play the greater part in this new, colder Hesperian era. In the same way that the chaoses formed from overpressured aquifers below the permafrost layer, the kilometres-deep trenches simply exposed the water-rich layers to the open air.

The sinking troughs were rapidly inundated and became repositories for debris washed or blown off Tharsis from both north and south. Then it was just a matter of how much lower the weight of that could drive the chasmata floors: any movement down would have made room for more material to be laid on top. A stack of sediment potentially 8 kilometres thick formed, representing the whole height of the deepest parts of Valles Marineris. The resulting lakes, deep and still, most likely periodically roofed with ice, ran the whole length of Valles Marineris, connected not by surface channels but by subsurface joints and porous rock.

Careful crater counting indicates that while the chasmata are early Hesperian in age, the outflow channels at the far eastern end, where the flood-scar of the Aurorae Chaos sits, are half a billion years younger, from towards the end of the era.

The top of the lake system was kilometres above the eastern lowland, with the sediment forming a natural dam. It's likely that water drained from the lakes by overtopping the dam, either continually, seasonally or during those longer, obliquity-inspired climatic variations. Over time an outflow channel, caused by erosion, would have formed. Every time water ran along it, the channel in the dam would have become deeper, wider and shorter, eroding backwards towards the lake.

Another day, another minor collapse. The channel opened wider and the new flow was stronger. The water tore harder at the banks and the bed of the channel, until suddenly the dam failed. All that water, penned up for so long, was suddenly free to move.

Most probably, like in the chaos-and-valles systems, flooding happened repeatedly, with the barriers failing intermittently and in stages, gradually moving the water level of the lake lower and lower. Much of the sediment that had collected in the lakes ran with the floods. There's evidence of layered deposits in the chasmata. Kilometres-high piles of lake-type sediments are still found on the floors of many: long ridges of isolated basement islands like the Eos Mensa, the Ius Mensa and the Coprates Mensa, or banked-up deposits near the chasm walls, categorically distinct from the rock type of the walls or the bottom of the chasmata. The soft sediment was mostly carried away, though. The amount excavated was typically heroic in volume – millions of cubic kilometres of stuff redistributed onto the northern plains of the Chryse Planitia.

The Aurorae Chaos sits below the outflow from the Valles Marineris, right in the firing line of the inland tsunami. Other chaoses lie along the path north of it – chaoses that might not have been formed by the upwelling of water underneath them, but by water tearing through.

We know what the Chryse Planitia looks like, intimately. Both Viking 1 and Mars Pathfinder – with its rover Sojourner – landed within its compass. It is an undulating plain with low hills and shallow depressions, and the ground is closely dotted with loose black rocks that range from shards to boulders, all interspersed with red soil. It looks a mess. The rocks have not been smoothed over time by steadily flowing water, but are rugged and chipped from frost. They are unsorted, dropped and abandoned. This is not a gentle terrain, but scuffed and wind-seared. A mountain has been thrown down here: lava blocks on an outwash plain, a memory of floods that once crossed the land and were then stilled.

OLYMPUS MONS

Olympus Mons is the largest volcano in the solar system and the second tallest mountain. It was the bright spot that nineteenth-century astronomers named Nix Olympica (the snows of Olympus), meaning that it's large enough to have been spotted from Earth using nineteenth-century telescopes.

Olympus Mons bears all the hallmarks of a shield volcano: this is a volcano with a roughly circular footprint, low-angled slopes running from the base to the summit and a wide, open crater – or caldera – at the top, so that it resembles a round shield dropped on the ground. Also typical of a shield volcano, Olympus Mons has lava flows and fronts down its flanks, collapsed lava tubes and, where the ground has been excavated by impact craters, exposed layers of solidified lava. The largest shield volcano on Earth is the Hawaiian island of Mauna Kea, which is 10 kilometres tall from seabed to summit. Olympus Mons is on a different scale altogether.

Olympus Mons rises 21 kilometres above datum, and because it sits on a below-datum plain to the west of the Tharsis Rise, we can add another kilometre to its actual height. East to west, it's 640 kilometres across. North to south, it measures 840 kilometres. As we'll discover, it used to be even larger. The slightly off-centre summit – which tends to the south-east – has six identifiable calderas that nest inside each other at the top, over an area roughly 80 by 60 kilometres. Lava overflowing the summit of Olympus Mons would have had to travel almost 400 kilometres before it reached the base of the volcano, staying liquid as it moved down the gentle gradient, pushed from behind by more erupting lava.

So how did Olympus Mons form? At some point in the early Hesperian, a hot spot in the mantle melted the crust beneath the Amazonis Planitia, and lava first came to the surface. The area where this happened is close to Tharsis, but close is a relative term – the three volcanoes that make up the Tharsis Montes are over a thousand kilometres to the east. But something definitely happened deep below this part of Mars that built this vast edifice rapidly. Most of Olympus Mons was complete within a few hundred million years, still within the Hesperian.

The reasons for believing this speed of construction are a little tentative, but they are compelling. The Hesperian was the high point of Martian volcanism, and some surfaces, hidden away on the extremities of Olympus Mons's long flanks, most likely date to the Hesperian. While the rest of the volcano is now veneered with later lavas, the core of it still sits inside.

The two most obvious features of Olympus Mons, after its sheer height, width and breadth have been considered, are these: firstly, the cliff that circles its base and secondly, the ring of debris that extends hundreds of kilometres out from there in obvious, discrete flows. Such features are not usual for Martian shield volcanoes, and they need explaining.

Shield volcanoes are formed by the flow of very fluid lavas. This defines them – it gives them their shape, their internal structure and determines how they interact with the encircling landscape. Lava is supposed to run down the flanks of the volcano from the summit and from vents in its side as rivers, forming lava tubes and levees that initially confine the flow and then break out later into broad fronts that spread onto the encircling plain. This is how a shield volcano slowly increases both its height and its diameter without substantially altering its overall shape. Shield volcanoes are supposed to have low-angled slopes of one or two degrees at the point where they meet the surrounding land. There definitely shouldn't be a cliff, let alone one that is, in places, 5 kilometres high.

Then there's the debris ring, usually referred to as the aureole, which stretches out in all directions from Olympus Mons's base; north and west across the Amazonis Planitia, south towards the Dichotomy and east in the direction of Tharsis. There are several distinct flows of material that form long fields of rubbly terrain, resting behind vast arced fronts of piled-up, thrown-down rock.

The obvious conclusion is that parts of Olympus Mons have broken off. The falling material formed giant landslides, which swept across the plains and ended up lying where they spilled, either stretched out on the lowlands or banked against local ridges. That conclusion encompasses the idea that the edge of the volcano was originally some 100–200 kilometres further out than its current cliff-faced position.

But there are immediate problems with such a scenario: kilometre-thick layered lava flows are really very solid and not prone to spontaneous collapse, especially when the deposits are at angles barely above one degree from the horizontal. Also, there's no known mechanism for a landslide to travel between 200 and

700 kilometres from its starting point – even factoring in Mars's reduced gravity, lower air pressure and every other conceivable environmental component. It simply couldn't happen.

And yet, the obvious conclusion is somehow correct. If we gather up all the landslides, trace them back to their beginnings and pack them up against the cliffs, we can make a perfectly shaped shield volcano out of them. Our assumptions must be wrong, somewhere.

They are. Detailed pictures of the cliff faces don't show serried banks of black lava flows edge on, but hard ridges of lavas interleaved with soft, grey, undifferentiated rock that avalanches dust. The angle of the cliffs isn't quite as sharp as expected either: rather than being near vertical, the average slope is somewhere between twenty and thirty degrees. This is steep, but certainly not as steep as could be made from solid blocks of basalt. The indication is, then, that the outer reaches of Olympus Mons aren't constructed primarily from rugged rock, but from loose ash, which if stacked up high enough, might well become mechanically unstable.

Even then, a collapsing mass of light ash and pumice can hardly be expected to travel hundreds of kilometres in a single coherent flow, before slumping neatly down on the ground with such a well-defined border – let alone repeatedly, towards all points of the compass. It might stretch to tens of kilometres, but the friction between the components of the slide would be high, and the energy of the fall would soon dissipate.

However, going back to the earlier contention that Olympus Mons was mostly formed in the early Hesperian, we might have a way out. What if these were not dry, dusty landslides composed of rattling rock that might have arisen out of the failure of a 5-kilometre-high slope? What if the base of the volcano had actually been submerged in a Hesperian-age ocean, and the collapse, transport and deposition of the aureole had all taken place underwater?

This proposition is startling given that everything we thought we knew about both the longevity and the depth of the northern ocean suggests that a deep sea shouldn't have been there at that time. But if – *if* – Olympus Mons was surrounded on all sides by water, then the erosion of its ashy flanks by waves would explain the presence of cliffs, and it would also explain the size and shape of the aureole.

Underwater landslides are of an entirely different quality to those in air. The water turns the debris into a dense slurry that minimises contact between the solid materials within the flow, while the difference in density between the flow and the surrounding water also prolongs the distance it can travel by lubricating the gap between the solid material and the seabed. We know such things happen around Hawaii, where submarine slumps and debris flows are common to all the volcanic islands in the chain.

In context, then, Olympus Mons probably started as a submarine volcano on the sloping sea-floor west of the Tharsis Rise, sometime during the late Noachian or early Hesperian. It grew rapidly over the next ten to a hundred million years, breaching the surface of the sea on its way to its 21-kilometre height. While lavas flowed from the central crater, and probably also from vents on the flanks, the outer reaches of the volcano were largely composed of loose ash and other aerial volcanic debris.

Saturated with water, the edges of the volcano repeatedly failed, forming huge submarine landslides that travelled underwater for hundreds of kilometres, before running out of energy or banking up against submarine ridges. Each collapse was associated with a tremendous tsunami that echoed up to the pole and back. The scarred cliffs carved by the collapses formed scarp slopes that were twenty degrees steeper than either the aureole deposits or the volcano itself.

This scenario does require the northern ocean to have persisted for longer than we had thought. It does require the crater-counted ages of the aureole to be wrong – the rubbly plains around Olympus return later dates. But other explanations are difficult to swallow. Transporting that amount of material that far from its source without the aid of some kind of fluid is highly unlikely: wind will not do, and neither will purely volcanic processes. The deposits in the aureole didn't form in place: their shape and position show they were moved there from the flanks of Olympus. And surely we can't suggest a deep ocean that lasted even longer? To push it from the Noachian into the Hesperian is difficult enough.

We are left with a paradox: Hesperian-aged deep-water deposits in a sea that should have been both retreating and freezing by that time, from a shield volcano that seems to have an outer portion composed of ash instead of lava that dates – using our most reliable technique – to an even later time. Such is Mars. We acknowledge its mysteries and move on.

ELYSIUM

Tharsis isn't the only volcanic province on Mars. It would be simpler if it was – then we would have one mantle plume that created the Great Dichotomy, and then wandered and created Tharsis. Yes, volcanism also occurred in and around large craters, where the thinned crust rose and the drop in pressure at the boundary with the mantle caused spontaneous melting, but much of Mars is underlain by flood basalts that appear to have no connection with either the craters or the plumes, and we simply have to accept that other mechanisms – ones that relied on very different conditions – were always in play.

The other major volcanic province is Elysium, where three volcanoes – one mons and two tholi – rise out of the Elysium plain in stark contrast to the low relief that surrounds them. Situated to the north of the Dichotomy, the largest volcano in the group, Elysium Mons, rises 12 kilometres above datum from a floor that is 2 kilometres below it. To the north is Hecates Tholus, a mere 8 kilometres tall, and to the south is Albor Tholus, only 6 kilometres high.

Elysium Mons is closer to what we think a volcano should look like. Its upper slopes are comparatively steep, but despite that, the lower slopes are so shallow that there's disagreement as to where the volcano ends and the surrounding plains begin.

Still, at some 400 kilometres across, Elysium Mons is a beast of a shield volcano, topped by the usual nested caldera that indicate several collapses – widely spaced apart in time – of the magma chamber below the peak.

Looking at the map, there's a sinuous line waiting to be drawn from Hecates, through Elysium and Albor, continuing southwards through the Cerberus region where flood basalts poured from rifts and barely raised small shield volcanoes, to Apollinaris Mons, sitting directly on the Dichotomy, and then further into the highlands to connect with Apollinaris Tholus and Zephyria Tholus, a low-relief shield volcano sitting on ancient Noachian rock. Making a connection between these volcanic centres is tempting and leads us to see a pattern of northward drift by some rogue plume fragment.

From all we can ascertain, though, there was no grand cause. Elysium began erupting in the very late Noachian or shortly after the transition to the Hesperian. Hecates started some half a billion years later and Albor a billion years after that. The eruptions on Cerberus look much, much younger. Just as Tharsis has no easy south-to-north ageing pattern, neither does Elysium, and we can only acknowledge that the order of eruptions was, if not arbitrary, then at least controlled by processes we do not yet understand.

Once again, we have to consider the extraordinary longevity of volcanism on a world where plate tectonics did not operate to create, move or destroy crust. Elysium Mons was mostly completed, built from the ground up, within a few hundred million years of the first eruption, and it still continued to erupt sporadically for the next 1.5 billion years. For whatever reason,

Elysium's molten rock stayed under the Elysium region for four billion years.

However, the power of the Elysium generator was an order of magnitude less than that of Tharsis. The Elysium region rose up from the surrounding plain, but it doesn't have the weight that Tharsis has. Elysium Mons is a significant volcano in its own right, but it's just under half as high as Olympus Mons. Tharsis dominates.

One property that volcanoes in the Elysium region do seem to possess is that at least some of the eruptions there weren't comprised of fluid lavas pouring from shallow cones and fissures to flood the surrounding terrain, but were composed of explosive, volatile-rich magmas that created significant quantities of fine ash, and with that, scalding-hot clouds of volcanic debris called pyroclastic flows.

Lavas that are more viscous, or sticky, keep their gases entrained for longer: as they rise up towards the surface, the sudden drop in external pressure causes the gas bubbles in the lava to abruptly expand, forcing it to foam and spray. This lava spray turns into the microscopic fragments of volcanic rock we know as ash; larger globules turn into frothy pumice. The eruption itself takes the form of a heavy, fiery blanket of gas, ash and rock that rolls down the flanks of the volcano. Often following existing valleys or folds in the landscape, these types of pyroclastic flow can drive on for tens of kilometres, and when they collapse they can add metres' worth of material to the ground in a single pass.

Other scenarios are even more violent. If the viscous lava forms a plug in the vent, pressure can build up behind it, and when mechanical failure occurs – either of the plug or of the containing vent – then the resulting surge of debris explodes outwards and can lay down huge quantities of rock, much of it gouged out from the existing volcano.

The steepness of the upper slopes of the Elysium volcanoes appears to indicate that the lava here was stickier than at Tharsis, at least some of the time, and that this would have been associated with both pyroclastic flows and voluminous ash production. In fact, volcanic ash plays a major role in one of the Hesperian's most significant geological formations – one that directly affects how we see Mars today.

THE MEDUSAE FOSSAE FORMATION

The one thing that we know about Mars – that we've always known – is that it's red. Since the first people looked up at the sky, since we constructed the first names in the first languages, Mars has always been the planet of fire, war, destruction and death, and it has been associated with gods that have those aspects. In Sanskrit, it was called Angaraka; in Egyptian, Her Deshur; in Hebrew, Ma'adim – all these names mean 'red'.

But Mars is not inherently red. Much of the rock that covers the surface is black basalt. The sediments that form from the erosion of this rock are mostly grey. The ice caps are white. Yet the colour of Mars is the colour of rust, from a sandy ochre through deep browns and reds to a dust-laden, baby pink sky. Where did this redness, this rustiness, come from?

The obvious answer – that it is rust – is both right and wrong. We can identify the chemical compound that's responsible for the redness, but also know it couldn't have been formed by the usual method.

Rust is the reaction of iron in the presence of water and oxygen to form iron hydroxides, which then desiccate to become iron oxides. Iron oxides are brittle compared with iron, and because they expand away from the surface of their parent metal, moisture and air can creep in through the outer layers of rust and corrode beneath them. While there's no native iron in Mars's crust or mantle – the thorough melting of Mars shortly after its formation sequestered all the metal iron in the core – iron-bearing minerals, silicates and oxides containing iron, form part of the usual suite of crystals found in both the crust and the mantle.

These iron minerals can break down in wet, oxygen-rich environments, gain more oxygen and turn into haematite. This transition is important, because while most iron minerals are black, haematite is red.

However, although Mars was wet at times, it was never – not once – oxygen rich. In dry climates where there's plenty of oxygen, only a thin veneer of rust can form, as it's water that opens up the chemical pathways and allows the easy transport of reagents. But in wet climates where there's no oxygen, no reaction can occur because the basic building-block of oxidation is not present. Without atmospheric oxygen, there's no oxygen dissolved in the water. The usual mechanism for converting black iron minerals into red haematite is simply not available.

Yet there's demonstrably red haematite on Mars, invariably as a component of the ubiquitous dust that blows around the entire planet. If we can talk about a water cycle that circulates water from ocean to clouds to ground and back, then we can also talk about a Martian dust cycle that circulates dust. Not all of the dust is haematite – a substantial fraction of it is magnetite and other volcanic silicates – but almost all of the haematite is dust. It's this dust that coats highlands and lowlands, that layers within and on top of ice. This is what makes Mars red.

Even the sky is red. On planets with an atmosphere, the sky looks blue because the molecules that make up the air scatter the blue part of sunlight more effectively than the red part, diffusing the blue light across the whole of the sky. As the Sun sets, the low angle of lighting means the blue is scattered out into space, and we tend to see more red. At the opposite, extreme, scenario, where a planet has no atmosphere, the sky is always black, even at midday, as there's no air to scatter the light.

Mars has an atmosphere, albeit a thin one now. As a result, we'd expect the sky to be blue-black during the day. Instead, it's pink from dust so fine that it can be carried high by the Martian winds. Part of the effect of this is to add a red hue to everything else, whatever its original colour.

Knowing why Mars is red still leaves us with two problems: firstly, how does the dust become red when we've no obvious way of making haematite? And secondly, how is there so much dust? The two questions are linked – of course – but we need to deal with them in reverse order: dust first.

This brings us to the Medusae Fossae Formation, which is a deposit of rock hundreds of metres thick that sits on the lowland plains, just north of the Dichotomy, between Elysium in the west and the edge of the Tharsis Rise in the east. Precisely what the Medusae Fossae Formation consists of has been the subject of fierce debate for decades: is it lake sediments, or a stranded raft of lighter-than-water pumice, or old glacial deposits or a carbonate shelf that formed at the edge of a shallow sea? For a while, it was even thought to be a vast deposit of dust, caught in the lee of the Dichotomy slope – a dust trap to end all dust traps.

Careful analysis brought about a surprising conclusion: the Medusae Fossae Formation isn't where dust goes to die; it's where it's born. The chemical signature of the global dust is very similar to that of the exposed rock of the Medusae Fossae, with just the right proportions of sulphur and chlorine. Material is being eroded from there and then dispersed by the Martian winds to every part of Mars.

If that's the case, then the Medusae Fossae Formation has to have been significantly larger than it is now to account for the sheer amount of dust on Mars, and it should also show wind-eroded features, both ancient and modern. That is what we find. The landscape is rich with signs of wind erosion: not just dunes and ripples, but more sculptural formations such as yardangs – carved, steep-sided ridges, tens of metres high and kilometres long, which align with the predominant wind direction – and pedestal craters, where an impact has locally fused softer sediment together, creating a hard-to-erode circle around it.

Because the Medusae Fossae Formation appears to be composed of loosely cemented volcanic ash and pyroclastic material, its surface is being continually reworked. It exists as both a primary deposit, where it has been laid down and is now being excavated by the equatorial wind, and also as a secondary formation, a wind-blown layer of sediment that overlies much younger lavas. It's a fascinating, mobile mess of tearing down and building up, and it makes calculating a crater-counted age enormously difficult.

It's almost impossible to pinpoint the formation's origin, but there are volcanoes nearby. If it is made from ash, the fact that it's flanked by the two major volcanic regions of Mars isn't going to be a coincidence, and Apollinaris Mons sits centrally on the southern edge of the deposit. Yes, crater counting can give a falsely young age – the whole technique is based on crater preservation, not crater erasure – but the surfaces the formation sits on are late Noachian or early Hesperian, and Hesperian and later lavas lie both on top and underneath in interleaved layers. The early to middle Hesperian is a guess for when it was formed, but a reasonable one.

Why red, though? Ash is grey, magnetite is black. How did this dust change colour in the absence of oxygen and, increasingly, in the absence of water? It turns out that it's the wind itself that creates the haematite through a process of mechanical milling.

Sand and other fine grains, when blown by the wind, move through a process known as saltation: essentially, a grain flicks up from the loose surface, is carried a short distance and then lands again. During its flight, the grain might hit another, and when it lands it certainly will. Each tiny impact abrades both it and whatever it strikes. As the particles get smaller and smaller, and create more and more dust, the surface area of each particle compared with its size increases dramatically, and this gives greater and closer contact during collisions. It's during these collisions that chemistry can occur that would normally only happen at higher temperatures and in the presence of water.

Mill a mixture of nine parts quartz sand and one part magnetite in a carbon dioxide atmosphere for long enough and you'll end up with haematite dust. The exact mechanism for how this occurs is poorly understood, but it does work. Pick up enough of the Medusa Fossae Formation and carry it in the wind for a few billion years, and Mars will gradually, inevitably, turn red.

TRUE POLAR WANDER

We know that Mars's polar axis wobbles – that the angle of obliquity (the angle which determines where the poles are pointing) changes periodically and chaotically. However, if we took a globe of Mars, put a finger on each pole and spun it, whether we held it upright or at a steeper angle, our fingers would still be on the poles. The poles remain in the same place on the globe, and the globe always rotates about the same axis.

But true polar wander is where the axis of rotation itself moves – the poles physically shift. This isn't unheard of in general, and it isn't unheard of in the history of Mars either. It's already thought to have happened once, when the Great Dichotomy formed. That event was so long ago that no real evidence of it has managed to survive, but that the Dichotomy line ended up parallel to the equator – implying the poles moved to accommodate the change in the distribution of Mars's mass – is well established. How, then, is the Dichotomy now twenty degrees askew from that?

That this is the case is undeniable. The evidence is there in the differing depth of the crust, measured from the gravity map, and also in the line we can draw on the surface between the southern highlands and the northern plains. And then there's the belt of late Noachian and early Hesperian river systems that were formed in a band twenty-five degrees south of the equator

– the old equator, that is – fed by rain coming in off the northern ocean. Map out all those lines now, and they make sinuous, parallel paths around Mars, crossing the new equator and back. In order to straighten these inconsistencies out, the poles need to have shifted.

Where were the old poles? There are more than a few degrees of uncertainty here, but the north pole was probably located in the part of the Vastitas Borealis called the Scandia Colles, while the south pole was somewhere in the Malea Planum, to the south of Hellas crater. These are the places we need to look to for evidence that – for perhaps 1.5 billion years or so – these areas were covered, at least some of the time, by ice sheets.

So what do we find? The Scandia Colles is characterised by odd, hummocky terrain that's suggestive of both subsurface ice melting that collapsed the ground above, and also liquid water run-off from a retreating ice cap. The water would have gone deep underground, but there's still a significant amount of water-ice in the area. It's promising evidence, but not conclusive.

In the south, the Malea Planum is characterised by long ridges, valley systems and other features like muddy ejecta splashes from cratering events that suggest unusually saturated ground in an area that should have been dry, cold desert, given where it sits in the highlands. Again, not conclusive evidence, but what we find doesn't contradict our supposition that a pole used to lie here.

So if we decide that the poles did move, that Mars's axis of rotation did slide twenty degrees from where it was to where it is today, at some point between the formation of the rain-belt valley systems of wet Mars and present-day frozen Mars, how could it have happened? The answer is obviously Tharsis, but that's not the whole story.

The centre of mass of Tharsis, its point of balance, is below a point on the western flank of Pavonis Mons, the middle of the three huge Tharsis Montes volcanoes that dominate the west of Tharsis. This point now lies directly on the Martian equator, yet if the preceding information is true, then this same point was once twenty degrees north of the equator, before the poles moved.

Tharsis on its own has the same mass as the dwarf planet Ceres, and it covers over a quarter of the surface area of Mars. Something that huge could, and eventually would, unbalance the axis of rotation of Mars itself. At a certain point it grew so large that the planet's spin – its moment of inertia – became unsteady. There needed to be a correction, and that was best achieved by shifting the axis such that Tharsis's centre of mass was sitting exactly on the line of maximum rotational speed: the equator.

So far, so extraordinary. That it happened is agreed upon, and we've located a cause. All we need to do now is identify a mechanism to allow for it. But turning the weight of an entire planet against its existing moment of inertia isn't something that's easily accomplished.

One of the things we do know is that solid crust can be carried along, at a rate of millimetres a year, on a slowly moving mantle current. It happens on Earth; it might happen on Venus; it might have happened on early Mars. But this mechanism applies to rafts of crust broken into continent-sized slabs, with new crust being formed in one place and being dragged down and destroyed in another. By the Hesperian, we're certain that Mars's crust

– brittle, fractured to a depth of tens of kilometres, measuring 60 kilometres thick in the south and 30 in the north – acted as a single 'stagnant lid': a solid, continuous shell of light, cold rock enclosing the totality of the mantle and pierced only in very few places by volcanic eruptions.

Could the crust have moved as a single entity, hauling itself into its new position, not pulled by movements of the mantle, but driving itself using Tharsis as the counterweight and dragging the mantle along with it? It's a journey of over 1,100 kilometres between the current pole and the old pole, one-twentieth of the circumference of Mars. Over a billion millimetres. At a rate of two millimetres a year, it would have taken 500 million years to complete the process. No one said this was going to be fast – but it demonstrably had to happen.

Is there any other evidence we can find for this slow, inexorable rearrangement of the geography of Mars? Again, we're confronted by the problem that we're looking back billions of years into the past. Such a gradual turn would have minimised the stresses on the rigid crust, but it wouldn't have eliminated them completely. The Martian surface is scored with both extension cracks (graben valleys) and compression features (wrinkle ridges); not all of them are associated with Tharsis and its extra weight, nor can the remainder all be linked with the other volcanic provinces. Might they be evidence of the shifting crust? If we map them out and try to make sense of the patterns, it turns out that the results are patchy, confusing and contradictory.

There is, though, one feature on Mars I have already mentioned that might lend this story an extra layer of credibility. Take a look at the Tharsis Montes, the three principal volcanoes of eastern Tharsis. From south-west to north-east, these three huge shield volcanoes – Arsia Mons, which stands 12 kilometres above the surrounding plains, Pavonis Mons, 14 kilometres, and Ascraeus Mons, 15 kilometres – lie almost exactly in a line.

Three huge volcanoes in a line speaks of a common cause: a single plume of mantle material underneath that part of Tharsis. Either the plume moved, breaking through the thickened Tharsis crust each time, or Tharsis itself moved and the plume remained stationary. We have precedents for both: the single mantle plume that tracked towards the Dichotomy in the early Noachian; and the linear Hawaiian islands on Earth, shield volcanoes all, built successively over a stationary plume as the Pacific plate travels at a relatively brisk 9 centimetres a year across it.

At this distance, we cannot tell which was the cause of the Tharsis Montes; it is most likely a bit of both. But it does look extraordinary. The clearest indication would be a firm date for the first eruption from each volcano, but that data is denied to us, buried under kilometres of subsequent lavas. All we have to go on are tentative last dates, given to us by crater counting the collapsed summits of each volcano in turn, and this tells us only that volcanism persisted long after the Hesperian ended.

As with much of Mars's history, we are left with answers that may be wildly, extravagantly wrong. At least, though, we know what the questions are. Mars's poles of rotation have moved and we have a culprit, even if we cannot quite work out how they committed the act.

THE ICE CAPS

The move into the Hesperian brought physical and climatic changes at even the highest of latitudes. True polar wander was dragging the poles away from their previous positions, incrementally, into new territory. Because the planet was cooling, there were extensive periods of time when the temperature never rose above the freezing point of water, even at the equator. Mars's atmosphere was also thinning: water vapour was dropping out as snow and staying locked up as ice. Even the carbon dioxide was starting to solidify as temperatures at the poles plummeted. The paradox was that the lower the pressure became, the more likely carbon dioxide was to form a gas at any given temperature. Dry ice sublimated away as the pressure fell and formed a positive feedback loop that prevented both temperature and pressure from reaching absolute zero.

Ice caps formed – potentially irregularly and inconstantly – at Mars's poles. Periods of high obliquity exposed the poles to long summers and long winters, and seasonal snow and ice came and went across large swathes of Mars. In periods of low obliquity, it stayed uniformly cold at the poles all year round: snow accumulated there, pressed down on the layers beneath, and fluffy snow became hard ice. When a warmer climate returned, as it did chaotically for tens or hundreds of thousands of years at a time, the ice caps, both north and south, retreated.

The resumption of the cold climate reinflated the ice caps and they crept back across previously ceded ground.

But as the Hesperian progressed, the lack of new snowfall, the sublimation of carbon dioxide and true polar wander meant that the existing ice caps began to retreat in earnest. They did so unevenly, especially in the south, where there was more geography to interfere with the weather patterns. Cold sinks – areas which were unusually frigid for their latitude – sat between the pole and Tharsis, and also on the high ground between the Argyre and Hellas craters; the depths of the craters themselves, with their higher air pressure and lower elevation, were warm by comparison. The southern ice cap was always colder than its northern counterpart, due to its being some three to four kilometres above datum. It was simply colder at altitude when the atmosphere was relatively thick.

When an ice cap melts, it leaves behind footprints – telltale markers of interaction between ice, water and rock – and when we look at the Martian polar regions, we see these signs in abundance. In the north, there are underground ice lenses and surface collapse structures, especially in the Scandia Colles region, where the old pole is believed to have been. In the south, the Dorsa Argentea Formation is crossed and crossed again by features called eskers.

When water melts on the surface of an ice cap, it runs through crevasses and joints within the ice, widening them into tight tunnels which work their way down from the top of the ice cap, all the way to the bottom. Melting also occurs at the base of an ice cap, where it sits on the bare rock below; pressure and rising heat – ice is an excellent insulator – can cause hidden lakes to form, entirely covered by the ice above. These under-ice rivers and lakes meet, and the flow of water picks up sediment – ground-down rock flour, entrained dust, larger broken fragments lifted from the basement by the dragging of the ice as it flows under its own weight to the edges – and it makes its own debris too, using what it's already carrying to wear at the subglacial riverbeds.

Confined to their icy tunnels, these fast-flowing and voluminous rivers rinse through the ice caps until they emerge from the leading edge in great, frigid flows. There they spread out and drop their fine loads across sandy outwash plains. New river systems form outside the ice cap, allowing the water to continue on into basins near and far, until it settles and freezes solid.

If the ice cap retreats, the riverbeds are slowly exposed. Each ice tunnel melts down to just its banks on either side, which then disappear completely. The sediment that was confined within the tunnel settles on the bare rock, leaving a sinuous ridge on the surface, a bank of debris that marks the route of the river along the underside of the ice. These lines of sediment – these eskers – can be significant features, both in height and width, and also in how they print the landscape with an obvious climatic marker. If there are eskers present, winding their way across a plain, then there was once a melting ice sheet on top of it. And while eskers can be erased again by a resurgent ice sheet, they are remarkably persistent once they form – on Mars, erosion rates are low in the cold environment of the far north and south.

If the Dorsa Argentea Formation was once covered by ice, then the Argentea Planum was the temporary holding lake for the water on its northern edge. River channels from there cut across the Noachian highlands and snaked towards the locally deep low of the Argyre crater, where Hesperian-age sands and muds now cover the basin floor.

The ice caps continued to grow and fade in time with the astronomical calendar, but without replenishment their maximum extent would never be as great as it was at the end of the Hesperian. Mars's drying and cooling was surely taking a fixed course now, not to be diverted.

PART SIX

THE AMAZONIAN

(3 billion years ago to the present day)

A WORLD OF ICE

Of all the jobs you've been given, this has to be one of the strangest. You're back in your buggy, but you're not rolling across a flat, barren, arid landscape of dust and rock. Instead, you're looking up, up and even further up at a reclining wall of pale pink, lit by the low polar sun – a wall that you know isn't made of layered basalts or wind-blown ash, but of solid ice.

You can see the cold. The ice gradually – one molecule at a time – turns from dense solid to tenuous vapour. It collects as streamers of white smoke and tumbles in eddies down onto the ground, where it dissipates like a summer mist. You shiver, despite the fact that your spacesuit is at the perfect temperature, neither warm nor cool, and the fans circulating the air inside your helmet are simply ticking over. The nominal temperature outside is minus seventy, and yet you feel fine.

Up on the top of the ice cap, a full three kilometres above you, is a drilling rig that's boring out ice cores, ten metres at a time, as far down as the drill string will support it. Those cores are going to be stored in a freezer – a freezer on Mars, really? – and then analysed one by one by the resident glaciologist. She'll be looking at the fine annual layering of dust, melting small samples in order to retrieve the atmospheric gases and looking for any discontinuities in the record. She wants to fill in the timeline of climatic conditions over the last ten or twenty thousand years, to study it for variations measured in decades, centuries, millennia.

Your task couldn't be simpler, though. Find the oldest ice you can, use your portable rock drill to take some small cores and store them in the insulated box you've been given. The oldest ice is, of course, at the base of the ice cap. Usually, that would be the most difficult and inaccessible part, but here on Mars you can drive up canyons carved by the freezing polar winds as they pinwheel off the pole, and face off against bright walls of ancient ice deep in the heart of the polar cap. Even if the drilling rig can't get down this far, here's an easy way for the glaciologist to get the samples she needs.

It's a strange landscape. The floor of the canyon is covered with thick dust, released by the tonne from the ice cap, that has fallen to coat the ground and then gathered up again in shapes that range from ripples to mighty marching dunes large enough to swallow your buggy whole. The permafrost lies below this mobile, inconstant surface. Your tyres plough through the fine red carpet, stirring it up: it's heavy going, and there's kilometre after kilometre of it.

The walls of the canyon start as nothing: mere lobes of dust-covered ice, indistinguishable from dunes. But they rise up, over the foreshortened horizon and to either side, and eventually you're in a flat-bottomed valley, flanked by oddly carved hills made of banded ice.

You've gone as far as you can. The canyon ends in a blunt curve that soars above you, a high, sloping buttress. You look up and you can see patterns in it. Superimposed over the yearly signals of dark summer melt and bright winter accumulation are thousand-year wavelengths, where those same annual marks are crammed together to produce smudged stripes in the exposed ice face. They alternate with wide bands, where the water vapour froze in abundance for a cycle, and spread out the years in a broad reach.

There's nothing to tell you just how old the ice is at the very base of this cliff; all you know is that this is where you have to drill. Perhaps there are markers in there – records of celestial events like cosmic ray bursts, solar flares, meteorite strikes – but that's for the glaciologist to find. You park up, set out the drill and carry both it and the sample box up to the cliff.

You have to turn your suit lights on. They shine harshly in the shadowed canyon, making the ice refract and glitter. Even though it's opaque, you can't help but see shapes in it. They're not real, and for a moment you wonder if your oxygen mix is wrong, but you check and it's fine – besides, alarms would have sounded long before you noticed anything. You put your gauntleted hand against the ice – briefly – and realise that of all the things you've touched on Mars, this is probably the most relatable. While your ancestors were busy knapping flints and hunting aurochs across the chilly plains of Europe, chasing the edge of the retreating ice front, there was another world where the ice was advancing, accumulating on bare Martian soil winter after winter, as another random, wayward change in obliquity altered the climate for another protracted season of cold.

INTO THE AMAZONIAN

As we've progressed from the Noachian to the Hesperian, we've been able to plant markers in the timeline for major events, even if some of our placements were tentative at best and wildly wrong at worst. But for the climate-critical moments, we were faced not just with a choice between vaguely reasonable guesses, but with actively contradictory evidence that confounds any attempt at a coherent storyline.

The transition between the Hesperian and the Amazonian is even more in dispute. It happened around 3 billion years ago, give or take 500 million years. The age of the exemplar surface, the Amazonis Planitia, is difficult to calculate accurately by crater counting, not because there are too many craters, but because there are too few.

By the time the Hesperian eased into the Amazonian, impacts had become unusual events. Not unheard of, certainly, but none of the big, planet-altering asteroids were left in Mars-crossing orbits, and in order to create anything approaching accurate crater-counted ages, smaller and smaller craters have to be considered – and in order for that to happen, the photographs we work from have to be very high resolution.

For a Noachian-aged surface, craters less than 16 kilometres in diameter are barely worth counting at all. For an Amazonian-aged surface, it's unlikely that there would be even one crater that large to count, and in order to get any idea at all of a comparative age, craters down to 100 metres and smaller have to be recorded.

In fact, because of the very low rates of crater-making events, the younger a surface is, the more difficult it becomes to date. A geological unit with a small total area – like a late lava flow or a riverbed – and with no craters at all is obviously relatively fresh, but just how young is it? A hundred million years? Ten million years? A thousand years? Counter-intuitively, it's easier to date an older surface, simply because it will have accumulated a reasonable number of craters early on in its existence.

Time collapses in on itself. There were few spectacular events in the Amazonian. No new volcanoes. No giant craters. No great oceans. Mars entered a long, cold, sparse epoch filled with ice, dust and wind. It became a deep freeze, a vacuum flask, a desert. We can point to features that exist, but not know when or in which order they formed. We have to treat them as scattered Polaroids thrown down on the ground: we know they represent part of the whole, but not all of it.

When we started our Mars journey, we thought we already knew what the endpoint would be: our Mars. It has visible ice at the poles, significant reservoirs of permafrost under the soil and glaciers at all latitudes. The red dust that coats everything is blown about on weak winds and the rate at which it wears down exposed rock is less than a single millimetre every million years. Meteorites hit fitfully. Landslides millennia in the making tumble down slopes in clouds of dust, settle and gain the appearance of having been there forever. Marsquakes shake the ground, but do not stir it. Everything seems static and sterile compared with what has gone before.

We know, with some degree of certainty, that Hesperian Mars was an active planet, but the processes on Mars today are literally glacial, and so we ask ourselves, when did all this activity cease? Was it an abrupt cut-off or a gradual, terminal decline from warm and wet to cold and dry?

But while we've long been convinced that Mars is dead, that nothing has happened in the Amazonian except dust and decay, that view is now undergoing some radical revision due to a fresh – and better – understanding of the climate of Mars. We know that the obliquity of the poles changes chaotically. That the latest large swing from high obliquity to a now-moderate average of twenty-five degrees occurred only four million years ago. That low-latitude glaciers formed at obliquities of thirty-five degrees perhaps within the last two million years. The suspicion is growing that there were – and still might be – pulses that reignited the spark of Mars.

Were there seasonal lakes in the early parts of the Amazonian? Most likely, yes. Were there rivers? Again, there's now evidence for them where there wasn't before, thanks to higher-resolution pictures. Were these rivers simultaneous with glaciers and permafrost sculptures? Yes, and one was probably the result of the other melting. There were also eruptions of lava and ash from volcanoes and rifts that we can date not just to three billion, two billion or one billion years ago, but to one hundred million or even ten million years. Just how close to the present do we see these events occurring? Could they be happening now? Have we, in fact, simply caught Mars napping?

Potentially. The physics of an escaping atmosphere and a cooling core demands a winding-down of Martian processes, but there are good reasons to assume that, superimposed over that billions-of-years-long flattening curve, are eras of surprising vigour. We can curse ourselves that we can't see Mars in that state, but we can still scour the surface for signs of activity and imagine. Onwards.

THE AMAZONIAN CLIMATE

Mars's Amazonian climate is the result of a complex network of strings. Pull on one and they all move. No one factor 'controls' whether ice advances or retreats, whether water flows or is locked into solid form. The obliquity of the poles is the main driver determining the amount of solar heat at the surface, but we can overlay this with the eccentricity of Mars's orbit, which determines the distance of the planet from the Sun throughout the yearly cycle, and the precession of the equinoxes, the gradual movement of the timing of the summer and winter seasons during each orbit.

While some of these variations are predictable in the short term over a few million years, in the long term they are chaotic. We can't extrapolate back into the past from the present because we know we would be wrong, and there's no way of telling how wrong. We can give broad upper and lower bounds to the state of Mars in the past, but we simply don't know when any supposed temperature regime would have operated, or how long it would have lasted.

Modelling the planet's behaviour can help here: in the same way we use climate models of our own planet to determine patterns and trends and then extrapolate those into the future, we can also make climate models of Mars and extend those into the past. We know that high-obliquity angles drive the melting of polar ice caps and the growth of low-latitude glaciers. We know that low obliquity causes those glaciers to retreat and the ice caps to grow.

Certainly, the ice doesn't appear from nowhere – it's not magicked into existence, nor is it dispelled with a wave of the wand; it has to come from what's left of the Martian water cycle. But this water cycle bears no similarity to the one we see on Earth. The Amazonian sees water move from relatively warm to relatively cold regions, without ever becoming a liquid; it's a cycle of solid-state sublimation and deposition, with the vapour transported by the wind. There are caveats: there's significantly more accessible water-ice in the north than the south, because it appears that much of the highlands are mantled with a thin layer of dry, wind-blown dust that insulates subsurface ice. And because of the Dichotomy, winds tend not to cross from north to south.

If conditions become warmer at the poles during high-obliquity periods, they conversely become colder at lower latitudes. This might seem counter-intuitive, but whereas the poles receive half a year of continuous sunlight at a time, the lower latitudes are plunged into cold, dark night once every sol and the Sun is only overhead briefly during spring and autumn. During high-obliquity periods, the equatorial regions are cold. During low-obliquity periods, the polar regions are cold.

We also know there are places on Mars that are cold sinks: we've already noted one in the region between Tharsis and the south pole, and another on the high ground between the Argyre and Hellas craters. There are other places where local conditions overwhelm global ones, too. If we run the Martian climate models to take account for geography, we find more traps for cold air: the western flanks of the Tharsis Montes and Olympus Mons, the eastern portion of Elysium Mons and the eastern basin and rim of Hellas. And these aren't the only places where glacial features are found on Mars – there are also high-altitude equatorial glaciers on the flanks of the tall volcanoes and lower-latitude glacial activity in areas such as the Phlegra Montes, a finger of a mountain range that reaches north from Elysium and into the Vastitas Borealis.

A survey of ice-related features – signs of left-behind glacial debris that make patterns of lines and lobes and characteristic concentric slumps inside craters that show that ice once filled and then flowed there – tells a story of mid-latitude ubiquity. Between twenty-five and sixty degrees, both north and south, these markers are everywhere.

It seems that ice accumulation away from the poles is a recurring feature of Mars: every time the obliquity shifts to steeper angles, ice doesn't so much march towards the equator as get blown there, to sit inside relatively young, fresh, steep-sided craters that can still trap a denser, water-vapour-laden atmosphere, and on the leeside valleys of high ground. If we want to see what that might look like, we can gaze down at the Korolev crater, which is in the Vastitas Borealis.

At 80 kilometres across, Korolev has a steep, high rim, a low floor and a 60-kilometre-wide dome of pale water-ice that covers the centre of the basin to a depth of almost 2 kilometres – that's more than 2,000 cubic kilometres of ice. Because Mars is currently in a low-obliquity period, the sight of Korolev is start-ling and rare: the northernmost plains have few large craters, and conditions have to be just right for ice to accumulate there, even at seventy degrees north. Only a few million years ago, these pearls of ice, nestling inside their crater shells, would have been commonplace across the mid-latitudes, numbering in their hundreds. Korolev is both a survivor and a sign of another Mars.

The most important feature of the Amazonian is this: even though the astronomical variables have been chaotic throughout the entire period, they have repeated again and again over the last three billion years. The climate has veered from low-latitude glaciation to high-latitude icing and back, several times over, in whichever hundred-million-year window we look through.

This chaotic cycling between rapid change and lengthy stability has had profound effects on the landscape. The ground has swelled and shrunk. Rocks have shattered and slid. Slopes have slumped and crept. It's impossible to tell how much of history has been smoothed and stretched by the constant churn of the climate bands across the face of Mars.

EQUATORIAL ICE

Ice on Mars is both obvious – in those pale northern and south-ern polar caps – and hidden. The ice caps represent huge stores of water and also carbon dioxide, which can freeze out on top of the caps. Ice, though, is almost everywhere. Permafrost on Mars – subsurface ice bound inside the fractured subsurface – is found across the whole planet. The deeper layers, 7 or 8 kilometres thick, are found adjacent to the poles, but even the equator is frozen to a depth of 2 or 3 kilometres. It's true that this deep ice isn't going to interact with the atmosphere – it's not going to melt, or sublimate, or in any way affect the formation of snow or frost at the surface – but given the huge areas of permanently frozen soil, we can say that even though most of Mars's water is bound up deep underground, there's still plenty that's access-ible and close to the surface. Both sixty degrees north and south, the soil is usually at least half ice by volume. Taking a shovel to Mars is akin to digging into a dirty snowdrift. Ice is part of the ground, exposed in the northern plains and hiding under a thin layer of dirt in the southern highlands. We should, therefore, be able to see all kinds of landforms that are caused by, and associ-ated with, ice.

The iciest of features are the glaciers themselves. When we think of glaciers, we probably bring to mind a pale tongue of ice draped between the jaws of sharp grey mountains, lolling out towards a valley below or down to an iceberg-flecked sea. Rivers run out from the front of the glacier as it melts, and we imagine outwash plains, ridged eskers, deep blue lakes and moraines piled high with hard, fist-sized rocks.

This was once true of Mars, back in the more-temperate Hesperian. We call these glaciers wet-based, to describe the inter-action between the ice and the ground, which is lubricated with running water. A wet-based glacier can flow relatively quickly, moving as a single mass, gouging at the contact points with the rock at its base and sides as it moves. Debris is entrained within the ice from below, and this shows itself both in the output of sediment at the front of the glacier, and in the landforms it leaves behind when it retreats.

Not so the cold-based glacier. There's only the thinnest film of water between ice and rock here, created by the weight of the ice above. Mostly, the glacier cleaves to the ground below, deforming and flowing over it like slow wax on its way downhill. Consequently, there's no grinding at the base of the glacier, and there's no water discharge from the front. Debris collects on the glacier's upper surface from rocks falling from the valley or cra-ter sides, and it gets conveyed along, accumulating thickness until the ice itself is completely hidden by a broken cloak of shifting rubble. Bare ice, in a low-obliquity period like now, will eventually turn to vapour, but the cover on rock glaciers provides a substantial (albeit imperfect) barrier to the loss of ice.

These stealthy rock glaciers are difficult to spot because they appear to be simply extensions of the landscape around them: same colour, same roughness. But there are subtle tells that give them away. The first is their shape. With good radar data that pinpoints the height of objects to within a few centimetres, a rubble-filled valley floor that shows a characteristic convex bulge from wall to wall is likely to be filled with a rock glacier. In the same way, if we measure along the length of the valley, the profile will be that of a shallow slope downhill, terminated by a blunt nose. This is exactly what we'd expect of a glacier, if it wasn't covered by debris.

The patterns on the surface can also show what lies beneath. Because the front of the glacier tends to form a rounded lobe as it trickles like treacle downslope, the debris layer wrinkles front to back. These rounded ridges travel with the glacier, before tipping over at the nose and being subsumed. In the same way, the sides of the glacier pull out long lines of debris in the direction of flow. These curved and straight ridges, perched on the top of the glacier, are often the only visual clues to the ice below.

As a consequence of this distinctive protective mantling, we can find equatorial rock glaciers on the north-western flanks of the Tharsis Montes and Olympus Mons, and smaller ones across broad swathes of Mars, between thirty degrees north and thirty degrees south.

Ice-filled craters less than 20 kilometres across are abundant in these regions too, all hiding their ice beneath a rock-rubble casing. However, this ice has nowhere to go: there's no downhill within a high-rimmed crater, and so all it can do is slump down inside. The walls of the crater break with age, supplying the ponded glacier with more debris, and the results are series of concentric ridges inside the crater, on top of the ice deposit. Given the utilitarian name of concentric crater fill, this was recognised as a landform well before the mechanism that caused it was understood.

While smaller craters are more able to keep their ice, as they can more completely cover the ice core with a barrier of debris, larger craters – like Korolev – will lose their ice faster. They're likely to be shallower, with older, more eroded rims that leak both air currents and sunlight into their depths, and being larger, any debris cover they might acquire will be limited to the margins of the glacier. With the centre exposed, the ice will sublimate away to almost nothing over several million years, until only a sliver is left on the north-facing internal wall.

But what of the vast mid-latitude ice plains of the north, and the mantled highlands of the south? One of the most obvious and ubiquitous signs of ice-rich soils is the peculiar shape of pedestal craters. We've met these once before in the Medusae Fossae Formation; they're craters caused by meteorite impacts in the usual way, but the whole structure appears to be raised up above the surrounding land as a rough circular plateau encompassing an area larger than the ejecta blanket, with the crater itself at the centre. The entire plateau stands tens of metres above the local level and is delineated by a steep scarp. But pedestal craters can also be explained by ice.

Icy ground forms craters like any other solid. The ejecta – rock melt, water vapour and broken, liquid-water-rich ground – is blasted out of the crater and deposited around it; the thickness and breadth of the ejecta blanket is proportionate to the size of the crater. The crater settles and then time gets to work.

If the ice is unstable – that is, if the conditions are such that water-ice turns into vapour – the soil around the crater loses volume. The ground literally deflates, sinking lower than its original level. We might assume that the crater would also simply disappear as the ice vanishes. Starting from the rim and moving downward, it would be eaten away until all that would be left is a shallow depression to mark where it once was, and even that might vanish.

However, while the ground around the crater falls away, the crater itself and its immediate surroundings seem immune to the loss of ice. This is partly for the same reason that rock glaciers persist: the ice is covered in debris. But that cannot be the whole reason, because the protected zone seems to extend beyond the reach of the ejecta, and the crater itself is also well preserved. How pedestal craters appear to armour themselves against erosion is another Martian mystery; it's likely due to the shock of impact welding soft sediment together, but how that mechanism works in ice-saturated ground is neither clear nor understood.

What we do believe is that the height of the pedestal – the plinth on which the crater is embedded – has nothing to do with the size of the impact, but is solely a function of the erosion of the ice-rich soil layer. Some scarps are 250 metres high, but these are unusual; the much more common height of a pedestal is 10–20 metres.

Naturally, pedestal craters are not immune to the ravages of the ages. As the ice is lost from the margins, the scarp slope loses its integrity and becomes easy pickings for the wind, which nibbles away at it until the crater collapses and become unrecognisable. But so many pedestal craters are persistent from forty degrees north, or south, to the pole, they are a firm Amazonian feature – one that represents both the past and the future.

HIGH LATITUDE ICE

Closer to the poles, where the permafrost is thicker and ice component greater, we see other extraordinary landforms that we can categorise as periglacial – those that are created close to an ice sheet or glacier. The processes that make them are completely dominated by the presence and behaviour of ice, where it's laid down and where it's taken up.

Polygonal cracks are a maze of interlocking lines that look like the remains of a dried-out puddle, but on a much larger scale, and they're formed by intense cold, not drought. The top of a permafrost layer grows and shrinks with the seasons. In winter, the icy ground contracts, splitting at the surface and forming cracks into which pure water-ice frost can fall; in summer, the warmed ground expands, pushing against the icy wedges and compacting itself a little more. Come winter, the cracks reopen and the ice wedges grow again. This forms the tesserae landscape common in the high northern plains.

Alases are circular depressions formed by the collapse of the surface permafrost due to ice sublimation. Material often slumps down from the margins to the middle, creating a beach-like effect at the edges. If these are low points in the landscape, then pingos are local highs: ice-cored conical hills that swell above ground level on permafrost plains. As the ice inside expands, it heaves the soil upwards to produce what look like little, steep-sided volcanoes with cracked, baked-muffin tops. Pingos grow only slowly, but a fully developed one can be hundreds of metres across and tens of metres high after centuries. They can and do deflate, leaving a circular ridge of material surrounding a sunken basin – which looks very much like an impact crater, but without the ejecta.

Intriguingly, we know from our own permafrost regions on Earth that these processes are all enhanced by the presence of liquid water. Polygonal cracks fed by meltwater form their characteristic landscape faster and more completely. Alases often hold seasonal meltwater, and pingos might use that water as their core and form at the bottom of those same dried-up lakes. This brings in the possibility that the Amazonian, like the Hesperian before it, might not have been as cold and dry as we earlier believed. We know that the climate veers wildly, and water-ice is transferred from the poles to low latitudes and back again. Could water have been stable at the surface during Amazonian times, and could it be again?

It's true enough that there's no current evidence of liquid water on Mars. There are no lakes, no rivers, no seasonal streams or ponding of groundwater: even the dark streaks that seemed to ooze out of crater walls and valley sides are now believed not to be salty brines, but simply small-scale flows of fresh black sand, released from the frozen surface by warmer summer temperatures. However, it's also true that the current range of conditions on Mars is temporary.

There were almost certainly times during the Amazonian that liquid water did exist at the surface: there's evidence that shallow rivers appeared locally in the mid-latitudes, draining into craters and, more importantly, out of them, indicating that they were at some point full to the brim. Little deltas and sinuous valleys complemented the rivers. They're almost impossible to date, since they cover only a small area of the surface, and it's also difficult to estimate how long they persisted for, but if we were to guess, they would have been active somewhere between the late Hesperian and the early Amazonian, turning on and off and on again, with each period lasting for several million years. Snow-melt, rather than groundwater, was probably the driver, given the locations of these features. Discrete drainage systems, based on local highs and lows, flowed as long as there was water. Then they just stopped. They ponded, they evaporated or drained away, or they froze and then turned to vapour, potentially forming alases or pingos.

The Amazonian could well have cycled between long periods where the only variation was exactly how cold it was, and brief million-year windows where the temperature – and the pressure – crept up enough to allow free water at the surface, even while the deep permafrost remained intact. But there remains the potential that in the recent past, Mars turned seasonally, if unspectacularly, wet. No wholesale melting of ice, just enough for the conditions to tip over to allow some surface flow for a few days, or a month or two, over the summer.

To show just what a knife-edge these climate calculations are on, we only need to consider the discoveries of the Phoenix lander, which touched down in 2008 on the Vastitas Borealis, at sixty-eight degrees north. Aiming to operate during one northern hemisphere summer, Phoenix was deliberately designed to look for water and conduct water-based experiments. Previous surveys had suggested that there was ice within a metre of the surface of the polygonally cracked ground; in fact, as the rockets that dropped the lander safely down also blew the surface layer of dirt away, they revealed bright ice within a mere 5 centimetres of the top of the soil.

It was light at the landing site nearly all the time, as expected for that high latitude. The Sun rose and dipped during each sol, but never dropped below the horizon. Phoenix relied on solar panels for power, so it was able to function continuously, until the Martian winter came and the Sun slipped below the horizon. Phoenix succumbed to the cold and the dark after a remarkable 157 sols.

During its months of operation, Phoenix reported on everything its cameras could see and its sensors could touch. Temperatures rose to −20°C during the day and fell to −80°C at night. Thin fog formed close to the ground when the Sun was at its lowest, and frost covered the rocks, before sublimating away as the Sun rose higher. Phoenix could see clouds, too, at various heights: high in summer, but as winter approached they formed lower and it snowed – lightly, but still snow. At no point was liquid water ever seen, and the ice the lander exposed, either with its rockets or its robotic arm, was unstable at the surface and began to pit and vanish within days of exposure. But it was close, so close. Given a slight shift in conditions – if Mars's closest approach to the Sun had coincided with the northern summer, for example – the temperature would have scraped above zero. With a higher obliquity, there would have been a hundred days of that climate.

The conclusion – perhaps a startling one – is that we have seen hints of a completely different Mars that is actually still possible. This means that the ice caps, both northern and southern, aren't permanent. They are recent constructions, temporary and cyclical, changing not just with yearly variations, or growing and shrinking as the seasons turn, but being partly or wholly obliterated by larger changes. We need to look to the poles for evidence of whether any of this is more than just wishful thinking.

THE POLAR REGIONS

The Amazonian ice caps are different to the ones that were present in the Hesperian, and not just because they're in subtly different places. The Dorsa Argentea esker deposits we visited earlier are most likely the site of an old ice sheet, but we don't know which route the poles took, nor the date by which true polar wander was complete. We don't even know if its rate increased at all, or whether the direction of travel reversed at any point.

Neither do we know whether the Amazonian ice caps were once wet-based, like their earlier Hesperian incarnations. Looking at both the north and south poles now, and the complete absence of contemporary outwash plains and other water-related signs, everything seems to indicate that the current ice caps are cold-based. That might also have been the case in the past, even during periods of high obliquity, but we need to determine how old the current ice caps are if we're going to make any attempt at answering our climate questions.

Mars's ice caps – their presence or absence, their size and their persistence – are related to the complex interplay of astronomy, climate and resources. Obviously, there'd be no ice caps without water to make them, and even then, early Mars probably didn't have permanent ice because the atmosphere was too thick and held its heat too well, despite the planet following the same chaotic obliquity patterns. Ice would only have formed as the atmosphere changed and the pressure dropped: seasonal at first, then all year round. The records for those earlier ice shields have been lost, and we don't know when that record began to be preserved. Ice, by its very nature, is volatile. It can be carved into the most spectacular shapes, but when it's gone, it's gone. It leaves few permanent traces: much of its ephemera can be overwritten by the next winter's fresh snow.

However, we still need to look. The current Martian polar deposits both have a threefold structure: the elements of this are different at each pole, since the higher altitude of the south pole, and the trapped weather system around it – created by the combination of the Argyre and Hellas craters – gives it a persistently colder quality.

They both have a temporary, seasonal layer of frozen carbon dioxide, which forms only during the coldest part of the winter and vanishes quickly when temperatures rise. It forms as a transparent glaze a few metres thick, but it still represents a significant portion of the atmosphere: a third of the Martian atmosphere freezes out every winter, switching between the north and south poles.

Below that is the first proper layer: in the north, it's a thick residual ice cap of white water-ice that grows and shrinks from winter to summer, but persists all year round. In the south, it's mainly carbon dioxide, with some water-ice. Both ice caps are currently growing. The smaller southern one because carbon dioxide freezes with the water-ice and traps it by virtue of its extreme coldness. The northern ice cap because it's the main repository for sublimated mid-latitude ice. But overall, water is being seasonally transferred from the northern hemisphere to the south pole, slowly but inexorably. At some point in the future, the amount of water-ice there will exceed the carbon dioxide ice, trapping the supercold layer below the surface: some kind of balance will be reached and the transfer will stop.

Beneath the residual ice cap layer is a series of thinly stacked deposits that make up the greater part of both polar ice caps. These are great domes of interleaved water-ice and dust, cut by ravines and chasms that spiral out of the main mass in the direction of the wind – anticlockwise in the north, clockwise in the south.

The northern ice cap is roughly circular, 1,000 kilometres across and 3 kilometres high. A great curved scar cuts kilometres down through it, almost to the bedrock in one part, called the Chasma Boreale, and on the southern side of that the ice cap continues as the Gemina Lingula. It looks for all the world like a great Catherine wheel of ice.

The centre of the southern ice cap is actually offset from the south pole by 150 kilometres and, as a structure, it's less well defined. It sits uncomfortably half in, half out of a large crater, Prometheus, and its ice deposits are dirtier. It's also irregular: 3–4 kilometres thick at its maximum, it extends northward in places as a plateau of dust-and-rock-covered ice. Like the northern cap, it has large chasmata – not one, but several, Chasma Australe being the largest – and this gives rise to three distinct lobes protruding from the main cap: Australe Lingula, Ultima Lingula and Promethei Lingula.

These chasmata, both north and south, have nothing to do with ground movement or the passage of water. They are created by wind alone. They do, though, allow us to see the interior of the ice caps without resorting to drilling. Inside, we can see a kilometres-high ladder of narrow, alternating light and dark bands of ice, with the colour of each band depending on the thickness of dust entrained within it. The ice is solid and the chasmata are steep-sided, with broken debris congesting their basements.

There's another layer at the bottom of the polar piles: the northern one is simply called the north polar basal unit, and the southern one we have encountered before – the Dorsa Argentea Formation. The north polar basal unit is a much more sand-rich deposit, and thickly layered, as if it's been made from the sublimation of the layered deposits above it. Which it probably has: take away the ice and we're left with a pile of wind-blown dust where the ice cap would have been. Sand that has blown out of the basal unit forms fields of moving dunes around the polar cap – the adjacent Olympia Planum has the same construction as the basal unit, except it's almost completely covered by dunes. It's not a stretch to suggest that reducing the ice cap to almost nothing would allow these dunes to advance onto the pole itself.

The southern ice cap appears to be more stable – and certainly more persistent – than its northern twin, but there are good reasons to believe that over the length and breadth of the Amazonian, both ice caps have died and risen and died again. The northern ice cap may only be a few million years old at most, and potentially only a few thousand years – crater counting really doesn't work on a surface that renews itself yearly. The southern structure is older, anywhere between a hundred million and ten million years old, but these ages have to be acknowledged as just best guesses.

The number of mid-latitude ice ages within the last few million years can be easily counted in the tens. Throughout the three billion years of the Amazonian, that number rises dramatically, so the ice caps become flickering, ghost-like features, with the ice switching from the low- and mid-latitudes to the poles and back again countless times.

The amount of water-ice held in the polar regions is about a 20-metre global equivalent layer. If the ice caps disappear, the atmospheric pressure doubles. If the pressure increases, so does the temperature. As the Martian climate yaws from low-obliquity polar ice caps to high-obliquity low-latitude glaciers, here's the tantalising hope: when the ice caps are dispensed with, we might just see free water again on the surface of Mars.

THE DUST CYCLE

Dust is not the same as sand. What we call sand is simply the fraction that is driven before the wind. Sand saltates; it gets picked up and travels in a ballistic arc, striking the ground further on. Dust, on the other hand, becomes part of the wind, and only drops when the wind can no longer hold it up. In practical terms, since Mars's atmosphere is so thin, this refers to particles no larger than two-millionths of a millimetre across.

Despite this size barrier, dust is ubiquitous. There's dust in the layered deposits at the north pole, and it covers the ice cap at the south pole. There's dust on the wide open northern plains. There's dust mantling the southern highlands. There's dust on the flanks of volcanoes and in the centres of craters. It forms ripples and dunes. It's carried from one part of the globe to the other, and eventually everywhere.

The Martian weather system, like all weather systems, is chaotic, but with a wide enough lens, trends appear and we can make predictions. It turns out that Mars has a dust season. It starts suddenly, with the southern hemisphere spring. The surface layer of winter ice turns to vapour, and the dust is left sitting there, waiting, poised for the first gusty winds to pick it up and carry it aloft. The lack of airborne dust early in spring allows the Sun to heat the ground directly – dust devils, spirals of warmer air that rise from the rock, are a constant factor in loading dust into the atmosphere. But as the Sun climbs higher and spring turns to summer, the winds that turn around the southern mid-latitudes get stronger, and they pick up dust from wherever it can be found.

Then there are dust storms. Most years there are only minor storms, but every so often there's one that shrouds the entire planet. They start in places like the Thaumasia Planum on southern Tharsis or around the Hellas crater, with a great plume of dust rising up into the sky. Mostly they fizzle out within a few days – Mars is quick to anger and equally quick to quieten again – but at other times, in other years, the storm gains in intensity. The atmosphere turns opaquely ochre and sunlight heats the cloud, adding more and more energy to the storm. The dust cloud spreads out across the southern hemisphere. Because of Mars's low gravity, the Martian atmosphere is tall and the dust can be carried a long way up, but Mars's atmosphere is also thin, so as soon as it stops churning, the dust falls out. Constant movement is a necessity.

Seeing a dust cloud rise up around Hellas and expanding so that it wraps around the entire southern hemisphere isn't an uncommon experience. The big storms, though – the ones where the dust crosses the equator and encroaches on the north – happen only once every decade or longer. The surface vanishes from view – even Olympus Mons, the Tharsis Montes and Elysium. Mars becomes a smooth, rust-red ball with only the pale poles visible. It can stay this way for weeks, and then it clears.

If we know something about how the storms start, grow and persist, we know very little about how they end. They should, reasonably, last longer, driven by the energy of the Sun striking the dust-laden air and heating it to new heights of opacity. But they don't. The winds don't even seem to lessen. The dust just seems to get tired of being up and comes down instead.

To be in a Martian dust storm is a confusing thing. Even though the sky grows dark and the Sun turns into a shadow, the fineness of the dust makes it look more like fog than objects in motion. The wind-speed isn't so troubling – barely topping out at 100 kilometres an hour, storm force but nothing greater – and the thinness of the air is such that, if we were standing in the middle of the storm, we would experience no buffeting or risk losing our balance. We would see bands of dust advancing on us like a mountain of cloud, cutting our vision to a few metres, and we would hear a low susurrus of sound, a thousand whispers just on the other side of our spacesuit helmets.

Despite the size, longevity and thickness of a big Martian dust storm, there's very little lightning. Detailed and concentrated study has shown that it's simply not a regular feature on Mars. This disappointing and somewhat counter-intuitive finding can be explained by understanding what good conditions for lightning look like.

Lightning needs, firstly, a very large difference in electrical charge between the ground and the cloud, and secondly, a gap between the two that doesn't habitually conduct electricity. We can see how Mars lacks both of those requirements: the cloud is already at ground level, and every level above it, and the dust itself behaves like an electrically conducting fluid – each tiny particle of dust carries a minute part of the electric charge, and all of them are in motion, giving no opportunity for any great difference in charge to build up.

This last property gives the dust its most dangerous feature, though. Because it's so easily charged, and so small, it sticks to everything, coats everything, gets into everything and is incredibly difficult to dislodge. The static cling of Martian dust means it can find its way into gears and seals, wearing them down and causing them to fail. It reduces the efficiency of solar panels. It contaminates samples and, because it's so very fine, if humans ever do reach Mars, it'll get into our artificial habitats and everywhere else, inevitably ending up as a feature of every Martian visitor's diet.

Even though dust is created slowly, the length of time that it has been accumulating – and being recycled – allows for deep deposits of wind-dropped dust. More than that, the dust can form rock: the electrostatic charge that individual particles carry can cause them to clump together so that they're no longer easily picked up. These duststones then form a stable surface and are eroded by the wind, not as loose cover but as soft rock, forming their own distinctive, wind-carved landforms. Duststone is the predominant rock type of the Amazonian, and the dust itself is a product of conditions that have persisted for three billion years. That's three billion years of material being stripped off the vast deposits of volcanic ash in the Medusae Fossae Formation, of sand grains grinding against each other and creating their strange mechanical–chemical interactions, of desiccated clays being lifted from crater floors and the bed of the old northern ocean, of a steady, thin patter of meteoric dust from space.

Just how much dust is there? The thickness in any particular region obviously depends on whether it's a net producer or accumulator of dust. For example, Arabia Terra, the only highland region that projects north of the equator, finds itself mantled in at least 20 metres of dust, and in places it can be as much as 60. Other regions clearly have nakedly exposed volcanic rock at the surface.

If a planet can have a water cycle, why not a dust cycle? Dust endlessly transported in a loop between low latitudes and high latitudes and back. The periodic cold, dry climate allows the equatorial regions to create the dust, while the higher-obliquity times release the reservoirs of trapped material in the polar caps. Some of it becomes stuck, some of it becomes free and there's always the wind to scour more from the volcanic source regions. If there's one thing Mars does well, it's making dust.

AMAZONIAN VOLCANISM

The start of the Amazonian was three billion years ago. No matter how much we try, we can't comprehend how distant that time is from us now. The earlier stages of the Amazonian – call it the first 500 million years or so – are further from us now than all but the very oldest terrestrial rocks, but that point marks the almost-end of our Martian journey. From the very beginning of Mars, when the hot mantle failed to set in motion a viable system of plate tectonics, we could have predicted that, after a frenetic start, all activity would tail off to a disappointing end.

So it's with a degree of discomfort that we learn that there were volcanoes still erupting in the Amazonian. Why that happened is one of those greater mysteries. Mars has had an extraordinarily persistent and active life, far beyond anything we could have expected. Because of its size and its history, by this period it should have been, if not cold, at least cooling like a casserole in a turned-off oven – one with a very solid lid.

So we appear to have a record of volcanic eruptions throughout the Amazonian, of the same very fluid lavas that we had in the Noachian when volcanoes first rose up on the rim of Hellas. Tharsis is coated with such lavas, as is Elysium, and there is little variation from the beginning until now.

Alba Mons is the northernmost of the Tharsis volcanoes, and

at first sight it's barely a volcano at all. It's fantastically flat for most of its construction, which stretches east to west for a full sixty degrees of longitude, almost 900 kilometres side to side. It's pretty much the size of France.

But Alba Mons appears to be two volcanoes, one on top of the other. The broad, near-flat apron of lava is surmounted by a more conventional shield volcano that is itself 360 kilometres across and 2 kilometres high. All this is balanced on the very edge of the Tharsis Rise, where it glides down into the northern plains. Surrounding it, and cutting through much of the earlier apron, are fields of linear tension faults – the Alba Fossae to the west and the Tantalus Fossae to the east both wrap around the dome-like summit, while to the south, the Ceraunius Fossae run steadily northward until they encounter the mass of the volcano, and falter against it.

The apron lavas are solidly Hesperian in age, but the summit is not, and neither are the two lobes of lava that extend east and west of it. The early Amazonian lavas were just as fluid as those that erupted in the Hesperian, but not quite as abundant. And even when Alba Mons appeared to be shutting down a few tens of millions of years later, there was still enough magma below for two small shield volcanoes to have appeared within the calderas of the collapsed summit.

Towards the middle of Tharsis, the three Tharsis Montes were also busy erupting. They continued to do so throughout the early and middle Amazonian, topping out their monumental structures at over 14 kilometres above the surrounding plain. And while their main building phase finished within a few hundred million years – itself an extraordinary length of time – they simply kept going. The youngest lavas on the flanks of Ascraeus Mons are barely twenty million years old.

Meanwhile, over on Elysium Planitia, between Elysium Mons and the on-the-Dichotomy Apollinaris Mons, potentially the youngest flood basalts on Mars occupy parts of the Athabasca Valles, with an estimated age of just two million years. In the planet's long history, this counts as yesterday.

If we take the suggested ages in these studies at face value, then there is one obvious, startling proposition we can make: volcanism is not yet finished on Mars. Mars appears to be still active in ways that we didn't think possible even a few years ago.

This is the enigma at the heart of our study. What are the chances of us sending spaceships to Mars at the precise moment that active volcanism – which has lasted for the best part of 4.5 billion years – ceases? We know that many, if not all, of the processes that have shaped the planet throughout its history, and certainly throughout the three billion years of the Amazonian, are continuing. The water cycle, powered by ice sublimation and deposition rather than water evaporation and condensation, is not just a yearly seasonal event, but one that extends over longer climatic periods and can melt the ice caps and allow free-flowing water on the surface. The dust cycle is intimately entwined with that movement and is a major source of new rock. And now there's the prospect that volcanism is not finished, and that we have simply had the misfortune to arrive in a quiet period.

Our view of Mars as cold, airless and dead is in need of serious revision. Its heart may be slowing, its lungs barely inflating, its blood sluggish, but there are planetary-scale systems still in motion. It might not be exactly what we want to see. For our own selfish reasons, we crave the idea of a second Earth – and to have one so close and yet so far is tantalising – but we know that Mars was never that. It was never ours. It is its own and we have to accept it for what it is.

PART SEVEN

THE
FUTURE

(The next 100–200 years)

WHAT CAN WE MAKE OF MARS?

Given that Mars isn't what we might want it to be, the question – unspoken, but present from the very first word – is what do we do with all this? Do we treat our exploration of Mars as an academic exercise, a worthwhile but ultimately esoteric study of a neighbouring planet that has both marked differences from and odd similarities to our own? Is it something we need to do in order to know our home better? Are we intending to use Mars as a first step, a proving ground, on our way to moving our nascent spacefaring civilisation out of the fragile basket that currently holds all our eggs? Or is it simply a place to exploit – a land-grab for pristine real estate, a new frontier that represents riches for a few or, less likely, for the many?

What we know of Mars has changed and is changing. The more robotic probes we send to trundle over the cracked lava flows and dusty sediments, the more detailed and accurate a picture we see: the only thing that's guaranteed is that there'll be more surprises along the way. But we already know enough to have a serious discussion about what Mars represents, as a laboratory for our research, a destination for our scientists and a place for us to colonise.

Our robotic ambassadors are still few and far between. Overall, two out of three Mars missions have failed – getting out of Earth orbit was particularly difficult early on – but we've been getting better. And while it's significantly easier to put a satellite into orbit around Mars than it is to put a lander on its surface, there have been some interesting mistakes there, too. For example, 1999's Mars Climate Orbiter was brought down when its thrusters were calibrated in imperial units but the on-board computer expected metric values.

Barring those missteps, getting to Mars orbit is simply a question of celestial mechanics and having enough propellant. Certainly, designing a mission, deciding what scientific instruments to put on it and how it can relay its information back to receiving stations on Earth are all difficult, technical problems: every kilogram of useful payload represents 100 kilograms of fuel required to get it to its destination. And while there are ways of helping ourselves to free energy – by waiting for preferential alignments between the planets and using gravity to slingshot our spaceship into position – it still comes down to how much grunt a launch system can provide.

We can put increasingly sophisticated sensors into orbit – including, but not limited to, cameras that use visible and infrared light, radar, lasers, spectrometers and magnetometers – and we can send probes down to the surface. These are sophisticated laboratories in their own right, whether they're static or on wheels, and in concert with their motherships, they represent the very pinnacle of our scientific and engineering capabilities. These landers have to survive the ferocity of launch, the months of space travel, the moment of separation from their carrier and the descent to the Martian surface – a descent with the entirely unironic name of 'seven minutes of terror'.

During this time, any and all difference between orbital height and speed and achieving a stationary landing on the surface needs to be accounted for, and there are no margins for error. One single slip will result in losing the ship, together with all the time and money spent on making it and getting it there and there's nothing anyone can do about it. Mars, at this point, is light-minutes away; any instruction we might send from Mission Control – to change an angle or the duration of a rocket burn, or even to abort the landing – will arrive long after the commitment to action has been made. The on-board computer can only do so much. Everything either happens close enough to the plan or we make a new crater on the surface.

A lander in low-Mars orbit is moving at between 3 and 4 kilometres per second. We need to slow it down in order to land, and we do that by throwing the lander at an angle at Mars. As it barrels at supersonic speeds through the upper reaches of the Martian atmosphere, it strikes the gas that makes up the atmosphere, transferring the energy of its motion to the atmosphere itself. Each individual impact with a gas molecule creates heat, and there are plenty of those, even in Mars's thin atmosphere. This means lots of heat – more than enough to destroy the lander – so there needs to be a heat shield between the working parts of the lander and the onrushing air.

But while there's enough atmosphere to cause us problems, there's not enough to help us. As soon as the lander is moving slowly enough that it's no longer at risk of burning up, we can deploy our parachutes and discard the heat shield. If the atmosphere were thick enough, we could float down to the ground from there. It's not. Even if we use huge parachutes, specially designed to work at very high speeds, then we still can't brake sufficiently hard not to plough into Mars.

To overcome this, we need to either absorb the inevitable impact or use another method to slow our descent to practically nothing. Most successfully, retro-rockets are used, either on the lander itself or as part of a sky-crane mechanism, where a framework hovers on spears of flame above the ground as it slowly winches the lander to the surface.

Each stage is fraught with danger. If one thing goes wrong, it all goes wrong. There can be no human intervention. Mars is a graveyard of our hopes and ambitions. Of the twelve landers to have made it as far as Mars orbit, fully half of them failed to get to the surface in good-enough working order, either burning up in flight or crashing into the ground – most recently in 2016, when a joint European and Russian lander called Schiaparelli misread an input and jettisoned its parachutes early. It was still 4 kilometres above the ground. The retro-rockets became very confused and the half-tonne lander made a new crater on the Meridiani Planum.

Of course, no one plans to fail: every lander sent so far has gone with the very best technology available at the time, but there are inevitably compromises that have to be made simply to get the spaceship there. It would be far safer for a descending craft to cut its orbital speed outside the atmosphere and coast to the ground in a controlled, powered descent, but the weight cost would be prohibitive, and it may be beyond our capability to deliver that much payload even into Earth orbit, let alone further away. We can argue that having a highly trained crew – a pilot, an engineer – on board any human mission to Mars will improve the odds of a successful landing significantly, but that's far from guaranteed.

This means that some of the people we send will die before they even take a step on Mars. It's inevitable, and we need to think about that. And the risks don't end with a successful landing. Mars is almost unconscionably far away – at best, 56 million kilometres or 3 light-minutes; at most, the other side of the Sun, 400 million kilometres. Messages – if they could be sent at all – would take more than 22 minutes to travel one way. Once on Mars, the crew would be on their own, with only the equipment and resources they brought with them or could scavenge from the environment.

They would have landed in an incredibly hostile environment. The atmosphere is both too thin and the wrong mixture of gases, rendering it unbreathable at any pressure. The soil is toxic, contaminated with chlorine-rich compounds called perchlorates at a level that is lethal to humans. With no Martian magnetic field to deflect, nor enough atmosphere to absorb, high-energy particles from the Sun, Mars is sleeted with radiation even on a normal day. If there's a solar flare, the radiation increases two hundredfold. Not everyone who goes to Mars and stays there will get cancer, but everyone who goes to Mars will run a much higher risk of getting cancer than they otherwise would.

The more insidious problem is that of the dust. We know that people who work in environments where they're exposed to ultrafine rock particles are going to breathe them in unless great care is taken and protective masks are worn. Silicosis is a grave threat to a person's health. More so are the associated cancers that go with certain types of rock dust. While outside, in a spacesuit, using tanked air, the dust would be safely away from an explorer's lungs. But if they brought it inside with them, the always-circulating air of the habitat would ensure it never settled and it would have to be deliberately removed – but these are the very smallest particles, so the filters would have to be the finest available.

The first humans on Mars would always be one step from disaster. A vital spare part, a chemical, a drug – all would be at the far end of the longest supply chain in history. That's the hard truth. But there's a very great deal we can do to make those risks as small as we can.

Firstly, we shouldn't go until we are ready. We already know that any Mars mission would need to get our explorers there, give them enough time on the surface to do useful science and then get them back again. We wouldn't be abandoning them there: we're not monsters. We'd plan everything carefully. We could pre-seed the landing site with everything the team would need, making sure it was all there before we committed to sending a crew. We could even land a return ship first, that would synthesise its own fuel from the Martian air – by splitting the carbon dioxide in the atmosphere into carbon monoxide and oxygen, or by adding hydrogen from Martian ice to create methane. And of course, the oxygen we farmed from the atmosphere or the ice for fuel could also be used to provide a breathable atmosphere, either of pure oxygen or buffered with nitrogen that we'd bring along to simulate Earth-normal conditions. As for water, it's too heavy to take with us, but we know we could just take a shovel to the ground, pressurise it and heat it to get all the water we would need.

The soil might be toxic, but we should be able to clean out the perchlorate, rendering it useful as a plant growth medium. The Sun would provide the bulk of our energy, although the sunlight is weaker on Mars than on Earth, and during dust storms that would reduce to almost nothing. A reliable backup would be needed: currently the only sources of that are highly radioactive and very heavy thermoelectric generators, but we do have them.

The radiation is dangerous for certain, but by burying our habitats under a thick layer of bulldozed soil, or by constructing them inside the entrance to a lava tube on the side of one of Mars's volcanoes, we could protect ourselves for the majority of the sol. For outside work, the astronauts would wear dosimeters to measure their exposure. Any Martian weather forecast would include a radiation rating, and there would be days when everyone would have to stay inside, no matter what.

Electric-powered rovers, assembled by the astronauts on the surface, would be simple enough to send ahead of time, as would the self-build habitats that the crew would create from kits. A successful Mars mission would see experiments in botany, geology and meteorology, as well as in the practicalities of growing food, harvesting water and air and surviving the harsh surface conditions. The architecture of any habitation would have to cope with wildly fluctuating temperatures between the very cold nights and the still-cold days, without leaking precious air. All these are engineering problems and not impossible to surmount.

To combat the dust, we'd have to think about washing spacesuits, vacuuming them off or trying to remove the dust using a static charge. Whichever solutions we chose, they'd need to be carried out regularly and rigorously to minimise the potentially fatal consequences of dust inhalation.

Beyond that? One of our chief concerns ought to be how to avoid contaminating Mars with our own bacteria. All landers – and all orbiting probes, but especially landers – are constructed in a surgically clean environment and undergo several rounds of decontamination, and even then the idea is only to minimise, not to completely eliminate, any microbial passengers. We could transport our terrestrial organisms to Mars and not realise it, but once we send a crewed mission, that scenario is inevitable. We cannot decontaminate a human being without killing them. We live in symbiosis with our bacterial load – without a panoply of flora both inside and out, we get very sick indeed. The consequence of contaminating Mars is that we could miss the signals of indigenous Martian life by masking it – or worse, exterminating it – with an invasive species from Earth. That would be a shame, since finding life is one of our principal reasons for going to Mars in the first place.

All of this is the context for scientific exploration. But what if we wanted to go and live there?

WE ARE THE MARTIANS

We realise that sending robotic probes and crews of scientists to Mars is both a worthy endeavour and a high-risk strategy. We can learn an enormous amount about another planet, and our own, by studying Mars in detail, even while we acknowledge that this will come at a cost: in terms of money, time, human life and the potential of ruining the environment we want to study.

Colonisation is a different matter. Permanent settlements on Mars would transpose our current economic model onto a new world, with all the exploitation of unrenewable resources and alteration of the landscape that this entails. We'd create buildings out of Mars material. We'd use Mars's atmosphere and water to make them habitable. We'd mine metals and engage in complex chemistry to form the hydrocarbons that are the rootstock of our plastics industry. We would, inevitably, produce waste that was seething with bacterial life.

Even if we started out with good intentions not to change Mars, we'd do it anyway. Humans are successful not because we fit in with our environment, but because we alter it to suit us. The high-stress situation we'd find ourselves in on Mars – no breathable air, no edible food, no sheltering structures, no potable water, no temperate climate – would require a technological, resource-heavy intervention.

The beautiful white pearl of the Korolev crater. Metal ores present in hydrothermal veins throughout Tharsis, Elysium and the highland volcanic provinces. Gold deposits in sedimentary sands on riverbeds and in ancient lakes. Once we arrived on Mars, the pressure to use them – even just a fraction of them – would be overwhelming. The cost of shipping everything we'd need from Earth would be huge, and we must never underestimate the satisfaction and the drive to gather resources from around us. The colonists will want to do this, not just to become less reliant on the old world, but actively to show their independence from it.

For the new Martians, being able to fix a broken valve, repair a split module or put a new wheel on a rover would all be matters of survival, not just contrariness. Even if there were a moratorium on strip-mining the water out of Korolev, if the choice was between dying and taking a pickaxe to the ice, the pickaxe would win. It's our nature.

This isn't to say that this scenario is intrinsically bad, but it's a debate that we have to have at the beginning of the process, not after it's under way. How do we want to treat Mars? Do we want to have a completely hands-off approach, which will preclude any of us ever planting a single ridged boot in the ochre dust of the Vastitas Borealis? Do we want to designate it as a planet purely for science? We have a model for that in our own Antarctic, where no country and no company can claim property or mineral rights on the entire continent, and where an international treaty oversees any conflicts that might arise. Or do we want to open Mars up to colonisation and exploitation, allow ownership on it and let those with the wherewithal to get there crown themselves kings?

There's a step beyond even colonisation that wouldn't just alter the planet, but change it utterly. The idea of terraforming Mars – making it more Earth-like – is a well-worn science-fiction trope, and as preposterous as it might at first seem, we have to have regard for the fact that we've been able to change the climate on our own planet in as little as two centuries by increasing the carbon dioxide content of the atmosphere. We already know that the Martian climate can be shifted using just one factor, the atmospheric pressure, and we can speculate on how we might boost that.

Most obviously, given our knowledge of how high-obliquity Mars sheds its ice caps, could we do something like that? Turning Mars would be beyond us, but deliberately heating the polar regions might not be. We could put large, thin, silvered mirrors in orbit and reflect sunlight downwards. If we melted the polar ice, we'd immediately double the current air pressure, but that would still put it at barely 1 per cent of the pressure at Earth's sea level. We'd need more gas from somewhere.

Dark objects absorb more heat than light ones. We could destroy Phobos and Deimos, both almost black, and scatter them across the water-rich northern plains to encourage them to defrost. Then there's comets, those leftovers from the solar system's formation, which regularly swing past the inner planets. If we could redirect them to strike Mars, then we'd inject fresh volatiles directly into the Martian atmospheric budget. We could head out to the tenuous trans-Neptunian storehouse of icy bodies and mine them, sending streams of precisely directed ice pellets inwards.

Perhaps you're now thinking, if we had the power and the technology to do all that, then we wouldn't need to live on Mars after all. We could use our expertise to build entirely artificial orbital habitats, or hollow out asteroids and pressurise them, rather than attempt to fundamentally transform Mars into something it could never really be: a home. There's a whole raft of difficult ethical questions to ask ourselves here, and the answers aren't obvious.

Mars has always been with us, that red point in the sky with a name in every language, freighted with meaning and endowed with supernatural powers. Mars is still there in our imaginations, even though our understanding of it has grown and grown, beyond the baleful influences of war gods, beyond the place from which three-legged death machines are launched across space in a green glow. It's become a solid reality that's no less wonderful and terrifying than those visions of our ancestors.

We cannot stand aside from the conversation to come – and it will come soon – as to what we do with Mars.

ACKNOWLEDGEMENTS

When I was first approached with this project, I distinctly remember telling my agent 'I will pay them to let me do this.' It didn't come to that, and in retrospect, asking a science fiction author who has a background in planetary geophysics to write a book on Mars wasn't such a random call after all – just one I never expected.

But my over-enthusiasm for highly detailed, specific, technical explanations had to be reined in somehow – and the four people who made this book even marginally readable need to be lauded for their patience, encouragement and clarity of thought and purpose. So, please – Antony Harwood, Simon Spanton, Pippa Crane, Clare Diston – take a bow.

BIBLIOGRAPHY

Abe, Y., Numaguti, A., Komatsu, G. & Kobayashi, Y. (2005) *Four climate regimes on a land planet with wet surface: Effects of obliquity change and implications for ancient Mars.* Icarus. 178, 27–39.

Anderson, R.C., Dohm, J.M., Golombek, M.P., Haldemann, A.F.C., Franklin, B.J., Tanaka, K.L., Lias, J. & Peer, B. (2001) *Primary centres and secondary concentrations of tectonic activity through time in the western hemisphere of Mars.* J. Geophys. Res. 106, E9, 20563–20585.

Andrews-Hanna, J.C. (2012) *The formation of Valles Marineris: 1. Tectonic architecture and the relative roles of extension and subsidence.* J. Geophys. Res. 117, E03006.

Andrews-Hanna, J.C. (2012) *The formation of Valles Marineris: 2. Stress focusing along the buried dichotomy boundary.* J. Geophys. Res. 117, E04009.

Andrews-Hanna, J.C. (2012) *The formation of Valles Marineris: 3. Trough formation through super-isostasy, stress, sedimentation, and subsidence.* J. Geophys. Res. 117, E06002.

Andrews-Hanna, J.C. & Phillips, R.J. (2007) *Hydrological modeling of outflow channels and chaos regions on Mars.* J. Geophys. Res. 112, E08001.

Andrews-Hanna, J.C., Zuber, M.T. & Banerdt, W.B. (2008) *The Borealis basin and the origin of the Martian crustal dichotomy.* Nature. 453, 1212–1215.

Baker, V.R. (2003) *Icy Martian mysteries.* Nature. 426, 779–780.

Baker, V.R. (2006) *Geomorphological Evidence for Water on Mars.* Elements. 2, 139–143.

Baker, V.R., Maruyama, S. & Dohm, J.M. (2007) Tharsis superplume and the geological evolution of early Mars. In: Yuen, D.A., Karato, S.I., Maruyama, S. & Windley, B.F. (eds.). *Superplumes.* Springer. pp. 507–522.

Banerdt, W.B. & Golombek, M.P. (2000) *Tectonics of the Tharsis regions of Mars: Insights from MGS topography and gravity*. Lunar Planet. Sci. Conf. XXXI, abstract 2038.

Basilevskaya, E.A., Neukum, G. & the HRSC Co-Investigator Team (2006) *The Olympus Volcano on Mars: Geometry and Characteristics of Lava Flows*. Solar System Research. 40(5), 375–383.

Basu, S., Richardson, M.I. & Wilson, R.J. (2004) *Simulation of the Martian dust cycle with the GFDL Mars GCM*. J. Geophys. Res. 109, E11006.

Bell, J.F., III, Savransky, D. & Wolff, M.J. (2006) *Chromaticity of the Martian sky as observed by the Mars Exploration Rover Pancam instruments*. J. Geophys. Res. 111, E12S05.

Bibring, J.-P., Langevin, Y., Mustard, J.F., Poulet, F., Arvidson, R., Gendrin, A., Gondet, B., Mangold, N., Pinet, P., Forget, F. & the OMEGA team (2006) *Global Mineralogical and Aqueous Mars History Derived from OMEGA/Mars Express Data*. Science. 312, 400–404.

Bouley, S., Baratoux, D., Paulien, N., Missenard, Y. & Saint-Bézar, B. (2018) *The revised tectonic history of Tharsis*. EPSL. 488, 126–133.

Bouley, S., Matsuyama, D.I., Forget, F., Séjourné, A., Turbet, M. & Costard, F. (2016) *Late Tharsis formation and implications for early Mars*. Nature. 531, 344–347.

Bridges, N.T. & Muhs, D.R. (2012) *Duststones on Mars: source, transport, deposition, and erosion*. Sedimentary Geology of Mars (SEPM Special Publication). 102, 169–182.

Bristow, T.F., Rampe, E.B., Achilles, C.N., Blake, D.F., Chipera, S.J., Craig, P., Crisp, J.A., Des Marais, D.J., Downs, R.T., Gellert, R., Grotzinger, J.P., Gupta, S., Hazen, R.M., Horgan, B., Hogancamp, J.V., Mangold, N., Mahaffy, P.R., McAdam, A.C., Ming, D.W., Morookian, J.M., Morris, R.V., Morrison, S.M., Treiman, A.H., Vaniman, D.T., Vasavada, A.R. & Yen, A.S. (2018) *Clay mineral diversity and abundance in sedimentary rocks of Gale crater, Mars*. Sci. Adv. 4. eaar3330.

Byrne, S. (2009) *The Polar Deposits of Mars*. Annu. Rev. Earth Planet. Sci. 37, 535–560.

Carr, M.H. & Head, J.W., III (2010) *Geologic history of Mars*. EPSL. 294, 185–203.

Carr, M.H. & Head, J.W., III (2019) *Mars: Formation and fate of a frozen Hesperian ocean*. Icarus. 319, 433–443.

Cassanelli, J.P. & Head, J.W. (2018) *Large-scale lava-ice interactions on Mars: Investigating its role during Late Amazonian Central Elysium Planitia volcanism and the formation of Athabasca Valles.* Planet. Space Sci. 158, 96–109.

Catling, D.C. (2014) Mars Atmosphere: History and Surface Interactions. In: Spohn, T., Breuer, D. & Johnson T.V. (eds.). *Encyclopedia of the Solar System (Third Edition).* Elsevier. pp. 343–357.

Chadwick, J., McGovern, P., Simpson, M. & Reeves, A. (2015) *Late Amazonian subsidence and magmatism of Olympus Mons, Mars.* J. Geophys. Res. Planets. 120, 1585–1595.

Chambers, J.E. (2007) Planet Formation. In: Davis, A.M. (ed.). *Treatise on Geochemistry, Volume 1.* Elsevier, pp. 1–17.

Clifford, S.M. (2001) *The Evolution of the Martian Hydrosphere: Implications for the Fate of a Primordial Ocean and the Current State of the Northern Plains.* Icarus. 154, 40–79.

Connelly, J.N., Bizzarro, M., Krot, A.N., Nordlund, Å., Wielandt, D. & Ivanova M.A. (2012) *The Absolute Chronology and Thermal Processing of Solids in the Solar Protoplanetary Disk.* Science. 338, 651–655.

Craddock, R.A. (2011) *Are Phobos and Deimos the result of a giant impact?* Icarus. 211, 1150–1161.

De Blasio, F.V. (2018) *The pristine shape of Olympus Mons on Mars and the subaqueous origin of its aureole deposits.* Icarus. 302, 44–61.

Dickson, J.L., Head, J.W. & Fassett, C.I. (2012) *Patterns of accumulation and flow of ice in the mid-latitudes of Mars during the Amazonian.* Icarus. 219, 723–732.

Dickson, J.L., Head, J.W. & Marchant, D.R. (2008) *Late Amazonian glaciation at the dichotomy boundary on Mars: Evidence for glacial thickness maxima and multiple glacial phases.* Geology. 36(5), 411–414.

Dickson, J.L., Head, J.W. & Marchant, D.R. (2010) *Kilometer-thick ice accumulation and glaciation in the northern mid-latitudes of Mars: Evidence for crater-filling events in the Late Amazonian at the Phlegra Montes.* EPSL. 294, 332–342.

Dohm, J.M., Baker, V.R., Maruyama, S. & Anderson R.C. (2007) Traits and evolution of the Tharsis superplume, Mars. In: Yuen, D.A., Karato, S.I., Maruyama, S. & Windley, B.F. (eds.). *Superplumes.* Springer. pp. 523–536.

Doute, S., Schmitt, B., Langevin, Y., Bibring, J.-P., Altieri, F., Bellucci, G., Gondet, B., Poulet, F. & the MEX OMEGA team (2007) *South Pole of Mars: Nature and composition of the icy terrains from Mars Express OMEGA observations.* Planet. Space Sci. 55, 113–133.

Dundas, C.M., McEwen, A.S., Chojnacki, M., Milazzo, M.P., Byrne, S., McElwaine, J.N. & Urso, A. (2017) *Granular flows at recurring slope lineae on Mars indicate a limited role for liquid water.* Nature geoscience. 10, 903–907.

Egea-Gonzalez, I., Jiménez-Díaz, A., Parro, L.M., Mansilla, F., Holmes, J.A., Lewis, S.R., Patel, M.R. & Ruiz, J. (2021) *Regional heat flow and subsurface temperature patterns at Elysium Planitia and Oxia Planum areas, Mars.* Icarus. 353, 113379.

Ehlmann, B.L., Mustard, J.F., Murchie, S.L., Bibring, J.-P., Meunier, A., Fraeman, A.A. & Langevin, Y. (2011) *Subsurface water and clay mineral formation during the early history of Mars.* Nature. 479, 53–60.

Forget, F., Haberle, R.M., Montmessin, F., Levrard, B. & Head, J.W. (2006) *Formation of Glaciers on Mars by Atmospheric Precipitation at High Obliquity.* Science. 311, 368–371.

Frey, H.V. (2006) *Impact constraints on, and a chronology for, major events in early Mars history.* J. Geophys. Res. 111, E08S91.

Gallagher, C., Balme, M., Soare, R. & Conway, S.J. (2018). *Formation and degradation of chaotic terrain in the Galaxias regions of Mars: implications for near-surface storage of ice.* Icarus. 309, 69–83.

Goetz, W., Madsen, M.B., Hviid, S.F., Gellert, R., Gunnlaugsson, H.P., Kinch, K.M., Klingelhöfer, G., Leer, K., Olsen, M. & the Athena Science Team. (2007) *The nature of Martian airborne dust. Indication of long-lasting dry periods on the surface of Mars.* Seventh International Conference on Mars. 3104.

Golombek, M.P. & Phillips, R.J. (2010) Mars tectonics. In: Watters, T.R. & Schultz, R.A. (eds.). *Planetary Tectonics.* Cambridge University Press. pp. 183–232.

Grant, J.A., Irwin, R.P., III, Wilson, S.A., Buczkowski, D. & Siebach K. (2011) *A lake in Uzboi Vallis and implications for Late Noachian–Early Hesperian climate on Mars.* Icarus. 212, 110–122.

Hargitai, H. & Kereszturi, Á. (eds.) (2014) *Encyclopedia of Planetary Landforms.* Springer.

Harrison, K.P. & Grimm R.E. (2008) *Multiple flooding events in Martian outflow channels.* J. Geophys. Res. 113, E02002.

Hartmann, W.K. (2005) *Martian cratering 8: Isochron refinement and the chronology of Mars.* Icarus. 174, 294–320.

Hartmann, W.K. & Neukum, G. (2001) *Cratering chronology and the evolution of Mars.* Space Science Reviews. 96, 165–194.

Hauber, E., Brož, P. & Jagert F. (2010) *Plains volcanism on Mars: ages and rheology of lavas.* Lunar Planet. Sci. Conf. XLI, abstract 1298.

Hauber, E., Platz, T., Reiss, D., Le Deit, L., Kleinhans, M.G., Marra, W.A., de Haas, T. & Carbonneau, P. (2013) *Asynchronous formation of Hesperian and Amazonian-aged deltas on Mars and implications for climate.* J. Geophys. Res. Planets. 118, 1529–1544.

Head, J.W., Greeley, R., Golombek, M.P., Hartmann, W.K., Hauber, E., Jaumann, R., Masson, P., Neukum, G., Nyquist, L.E. and Carr, M.H. (2001) *Geological Processes and Evolution.* Space Science Reviews. 96, 263–292.

Hynek, B.M., Beach, M. & Hoke, M.R.T. (2010) *Updated global map of Martian valley networks and implications for climate and hydrologic processes.* J. Geophys. Res. 115, E09008.

Hynek, B.M., Phillips, R.J. & Arvidson, R.E. (2003) *Explosive volcanism in the Tharsis region: Global evidence in the Martian geologic record.* J. Geophys. Res. 108(E9), 5111.

Hynek, B.M., Robbins S.J., Šrámek O. & Zhong, S.J. (2011) *Geological evidence for a migrating Tharsis plume on early Mars.* EPSL. 310, 327–333.

Hyodo, R., Genda, H., Charnoz, S. & Rosenblatt, P. (2017) *On the Impact Origin of Phobos and Deimos. I. Thermodynamic and Physical Aspects.* Astrophys. J. 845, 125.

Ivanov, M.A. & Head, J.W. (2006) *Alba Patera, Mars: Topography, structure, and evolution of a unique late Hesperian–early Amazonian shield volcano.* J. Geophys. Res. 111, E09003.

Johnson, C.L. & Phillips, R.J. (2005) *Evolution of the Tharsis region of Mars: insights from magnetic field observations.* EPSL. 230, 241–254.

Kadish, S.J., Barlow, N.G. & Head, J.W. (2009) *Latitude dependence of Martian pedestal craters: Evidence for a sublimation-driven formation mechanism.* J. Geophys. Res. 114, E10001.

Kadish, S.J., Head, J.W. & Barlow, N.G. (2010) *Pedestal crater heights on Mars: A proxy for the thicknesses of past, ice-rich, Amazonian deposits.* Icarus. 210, 92–101.

Kahre, M.A. & Haberle R.M. (2010) *Mars CO2 cycle: Effects of airborne dust and polar cap ice emissivity.* Icarus. 207, 648–653.

Kahre, M.A., Murphy, J.R., Haberle, R.M., Montmessin, F. & Schaeffer, J. (2005) *Simulating the Martian dust cycle with a finite surface dust reservoir.* Geophys. Res. Lett. 32, L20204.

Kallenbach, R., Geiss, J. & Hartmann, W.K. (eds.) (2001) *Chronology and Evolution of Mars: Proceedings of an ISSI Workshop, 10–14 April 2000, Bern, Switzerland.* Springer.

Kerber, L. & Head, J.W., III (2010) *The age of the Medusae Fossae Formation: Evidence of Hesperian emplacement from crater morphology, stratigraphy, and ancient lava contacts.* Icarus. 206, 669–684.

Kite, E.S., Matsuyama, I., Manga, M., Perron, J.T. & Mitrovica, J.X. (2009) *True Polar Wander driven by late-stage volcanism and the distribution of paleopolar deposits on Mars.* EPSL. 280, 254–267.

Kite, E.S., Sneed, J., Mayer, D.P. & Wilson, S.A. (2017) *Persistent or repeated surface habitability on Mars during the late Hesperian – Amazonian.* Geophys. Res. Lett. 44, 3991–3999.

Lammer, H., Odert, P., Leitzinger, M., Gröller, H., Güdel, M., Kislyakova, K.G., Khodachenko, M.L. & Hanslmeier, A. (2011) Origin and solar activity driven evolution of Mars' atmosphere. In: Okano, S., Kasaba, Y. & Misawa, H. (eds.). *Proceedings of the International Symposium on Planetary Science 2011.* Terrapub. pp. 13–31.

Laskar, J., Correia, A.C.M., Gastineau, M., Joutel, F., Levrarda, B. & Robutel P. (2004) *Long term evolution and chaotic diffusion of the insolation quantities of Mars.* Icarus. 170, 343–364.

Leask, H.J., Wilson, L. & Mitchell, K.L. (2006) *Formation of Aromatum Chaos, Mars: Morphological development as a result of volcano-ice interactions.* J. Geophys. Res. 111, E08071.

Leone, G. (2016) *Alignments of volcanic features in the southern hemisphere of Mars produced by migrating mantle plumes.* J. Volcanol. Geotherm. Res. 309, 78–95.

Levy, J.S., Fassett, C.I. & Head, J.W. (2016) *Enhanced erosion rates on Mars during Amazonian glaciation.* Icarus. 264, 213–219.

Mandt, K.E., de Silva, S.L., Zimbelman, J.R. & Crown, D.A. (2008) *Origin of the Medusae Fossae Formation, Mars: Insights from a synoptic approach.* J. Geophys. Res. 113, E12011.

Mangold, N., Ansan, V., Masson, P. & Vincendon, C. (2009) *Estimate of aeolian dust thickness in Arabia Terra, Mars: Implications of a thick mantle (>20 m) for hydrogen detection.* Géomorphologie: relief, processes, environment. 1, 23–32.

Mangold, N., Loizeau, D., Poulet, F., Ansan, V., Baratoux, D., LeMouelic, S., Bardintzeff, J.M., Platevoet, B., Toplis, M., Pinet, P., Masson, Ph., Bibring, J.P., Gondet, B., Langevin, Y. & Neukum, G. (2009) *Mineralogy of recent volcanic plains in the Tharsis region, Mars, and implications for platy-ridged flow composition.* EPSL. 194(3–4), 440–450.

Martín-Torres, F.J., Moyano-Cambero, C. & Trigo-Rodríguez, J. (2012) *Evolution of Mars Atmospheric Pressure and Temperature Modeling and Constraints from Meteorites.* Lunar and Planetary Science Conference. XLIII. Abstract 2840.

Massé, M., Le Mouélic, S., Bourgeois, O., Combe, J.-P., Le Deit, L., Sotin, C., Bibring, J.-P., Gondet, B. & Langevin, Y. (2008) *Mineralogical composition, structure, morphology, and geological history of Aram Chaos crater fill on Mars derived from OMEGA Mars Express data.* J. Geophys. Res. 113, E12006.

McCubbin, F.M., Smirnov, A., Nekvasil, H., Wang, J., Haur, E. & Lindsley, D.H. (2010) *Hydrous magmatism on Mars: A source of water for the surface and subsurface during the Amazonian.* EPSL. 292, 132–138.

Merrison, J.P., Gunnlaugsson, H.P., Knak Jensen, S. & Nørnberg, P. (2010) *Mineral alteration induced by sand transport: A source for the reddish color of Martian dust.* Icarus. 205, 716–718.

Michalski, J.R., Dobrea, E.Z.N., Niles, P.B. & Cuadros, J. (2017) *Ancient hydrothermal seafloor deposits in Eridania basin on Mars.* Nat. Commun. 8, 15978.

Mitrofanov, I.G., Zuber, M.T., Litvak, M.L., Demidov, N.E., Sanin, A.B., Boynton, W.V., Gilichinsky, D.A., Hamara, D., Kozyrev, A.S., Saunders, R.D., Smith, D.E. & Tretyakov, V.I. (2007) *Water ice permafrost on Mars: Layering structure and subsurface distribution according to HEND/Odyssey and MOLA/MGS data.* Geophys. Res. Lett. 34, L18102.

Montabone, L., Lewis, S.R. & Read, P.L. (2005) *Interannual variability of Martian dust storms in assimilation of several years of Mars global surveyor observations.* Advances in Space Research. 36, 2146–2155.

Morbidelli, A. & Raymond, S.N. (2016) *Challenges in planet formation.* J. Geophys. Res. Planets. 121, 1962–1980.

Mouginis-Mark, P. (2018) *Olympus Mons volcano, Mars: A photogeologic view and new insights.* Geochemistry. 78(4), 397–431.

Murchie, S.L., Thomas, P.C., Rivkin, A.S. & Chabot, N.L. (2015) Phobos and Deimos. In: Michel, P., DeMeo, F.E. & Bottke, W.F. (eds.). *Asteroids IV.* Univ. of Arizona. pp. 451–468.

Murray, J.B., Muller, J.-P., Neukum, G., Werner, S.C., van Gasselt, S., Hauber, E., Markiewicz, W.J., Head, J.W., III, Foing, B.H., Page, D., Mitchell, K.L., Portyankina, G. & The HRSC Co-Investigator Team (2005) *Evidence from the Mars Express High Resolution Stereo Camera for a frozen sea close to Mars' equator.* Nature. 434, 352–355.

Newman, C.E., Lewis, S.R., Read, P.L. & Forget, F. (2002) *Modeling the Martian dust cycle, 1, Representations of dust transport processes.* J. Geophys. Res. 107(E12), 5123.

Newman, C.E., Lewis, S.R., Read, P.L. & Forget, F. (2002) *Modeling the Martian dust cycle, 2, Multiannual radiatively active dust transport simulations.* J. Geophys. Res. 107(E12), 5124.

Ojha, L., Lewis, K., Karunatillake, S. & Schmidt, M. (2018). *The Medusae Fossae Formation as the single largest source of dust on Mars.* Nat. Commun. 9(1), 2867.

Osinski, G.R. & Pierazzo, E. (2013) Impact cratering: processes and products. In: Osinski, G.R. & Pierazzo, E. (eds). *Impact Cratering: Processes and Products.* Wiley-Blackwell. pp. 1–20.

Peters, S.I. & Christensen, P.R. (2017) *Flank vents and graben as indicators of Late Amazonian volcanotectonic activity on Olympus Mons.* J. Geophys. Res. Planets. 122, 501–523.

Pham, L.B.S., Karatekin, Ö. & Dehant, V. (2009) *Effects of Meteorite Impacts on the Atmospheric Evolution of Mars.* Astrobiology. 9(1), 45–54.

Phillips, R.J., Davis, B.J., Tanaka, K.L., Byrne, S., Mellon, M.T., Putzig, N.E., Haberle, R.M., Kahre, M.A., Campbell, B.A., Carter, L.M., Smith, I.B., Holt, J.W., Smrekar, S.E., Nunes, D.C., Plaut, J.J., Egan, A.F., Titus, T.N. & Seu, R. (2011) *Massive CO_2 Ice Deposits Sequestered in the South Polar Layered Deposits of Mars.* Science. 332, 838.

Phillips, R.J., Zuber, M.T., Smrekar, S.E., Mellon, M.T., Head, J.W., Tanaka, K.L., Putzig, N.E., Milkovich, S.M., Campbell, B.A., Plaut, J.J., Safaeinili, A., Seu, R., Biccari, D., Carter, L.M., Picardi, G., Orosei, R., Mohit, P.S., Heggy, E., Zurek, R.W., Egan, A.F., Giacomoni, E., Russo, F., Cutigni, M., Pettinelli, E., Holt, J.W., Leuschen, C.J. & Marinangeli, L. (2008) *Mars North Polar Deposits: Stratigraphy, Age, and Geodynamical Response.* Science. 320, 1182–1185.

Pierens, A. & Raymond, S.N. (2011) *Two phase, inward-then-outward migration of Jupiter and Saturn in the gaseous solar nebula.* A&A. 533, A131.

Platz, T. & Michael, G. (2011) *Eruption history of the Elysium Volcanic Centre, Mars.* Geophysical Research Abstracts. 13, EGU2011-13166.

Plescia, J.B. (2004) *Morphometric properties of Martian volcanoes.* J. Geophys. Res. 109, E03003.

Poulet, F., Bibring, J.-P., Mustard, J.F., Gendrin, A., Mangold, N., Langevin, Y., Arvidson, R.E., Gondet, B., Gomez, C. & the Omega Team (2005) *Phyllosilicates on Mars and implications for early Martian climate.* Nature. 438, 623–627.

Ramsdale, J.D. (2017). *Studies of Glacial and Periglacial Environments on Mars.* PhD thesis. The Open University.

Raymond S.N. & Morbidelli A. (2014) The Grand Tack model: a critical review. In: Z. Knežević & A. Lemaitre (eds.). *Complex Planetary Systems.* Proceedings IAU Symposium. 310, 194–203.

Richardson, M.I. & Wilson, R.J. (2002) *Investigation of the nature and stability of the Martian seasonal water cycle with a general circulation model.* J. Geophys. Res. 107(E5), 5031.

Rickman, H., Błęcka, M.I., Gurgurewicz, J., Jørgensen, U.G., Słaby, E., Szutowicz, S. & Zalewska, N. (2019) *Water in the history of Mars: an assessment.* Planet. Space Sci. 166, 70–89.

Robbins, S.J., Di Achille, G. & Hynek, B.M. (2011) *The Volcanic History of Mars: High-Resolution Crater-Based Studies of the Calderas of 20 Volcanoes.* Icarus. 211, 1179–1203.

Roberts, J.H. & Zhong, S. (2006) *Degree-1 convection in the Martian mantle and the origin of the hemispheric dichotomy.* J. Geophys. Res. 111, E06013.

Scanlon, K.E., Head, J.W., Fastook, J.L. & Wordsworth, R.D. (2018) *The Dorsa Argentea Formation and the Noachian-Hesperian climate transition.* Icarus. 299, 339–363.

Scanlon, K.E., Head, J.W. & Marchant, D.R. (2015) *Volcanism-induced, local wet-based glacial conditions recorded in the Late Amazonian Arsia Mons tropical mountain glacier deposits*. Icarus. 250, 18–31.

Schmedemann, N., Michael, G., Ivanov, B.A., Murray, J.B. & Neukum, G. (2014) *The age of Phobos and its largest crater, Stickney*. Planet. Space Sci. 102, 152–163.

Schorghofer, N. (2008) *Temperature response of Mars to Milankovitch cycles*. Geophys. Res. Lett. 35, L18201.

Séjourné, A., Costard, F., Gargani, J., Soare, R.J. & Marmo, C. (2012) *Evidence of an eolian ice-rich and stratified permafrost in Utopia Planitia, Mars*. Planet. Space Sci. 60, 248–254.

Shean, D.E., Head, J.W. & Marchant, D.R. (2005) *Origin and evolution of a cold-based tropical mountain glacier on Mars: The Pavonis Mons fan-shaped deposit*. J. Geophys. Res., 110, E05001.

Skok, J.R., Mustard, J.F., Murchie, S.L., Wyatt, M.B. & Ehlmann, B.L. (2010) *Spectrally distinct ejecta in Syrtis Major, Mars: Evidence for environmental change at the Hesperian-Amazonian boundary*. J. Geophys. Res. 115, E00D14.

Smith, I.B., Putzig, N.E., Holt, J.W. & Phillips, R.J. (2016) *An ice age recorded in the polar deposits of Mars*. Science. 352(6289), 1075–1078.

Smith, P.H. (2009) *Water at the Phoenix landing site*. Dissertation. The University of Arizona.

Smith, P.H., Tamppari, L.K., Arvidson, R.E., Bass, D., Blaney, D., Boynton, W.V., Carswell, A., Catling, D.C., Clark, B.C., Duck, T., DeJong, E., Fisher, D., Goetz, W., Gunnlaugsson, H.P., Hecht, M.H., Hipkin, V., Hoffman, J., Hviid, S.F., Keller, H.U., Kounaves, S.P., Lange, C.F., Lemmon, M.T., Madsen, M.B., Markiewicz, W.J., Marshall, J., McKay, C.P., Mellon, M.T., Ming, D.W., Morris, R.V., Pike, W.T., Renno, N., Staufer, U., Stoker, C., Taylor, P., Whiteway, J.A. & Zent, A.P. (2009) *H_2O at the Phoenix Landing Site*. Science. 325, 58–61.

Soare, R.J., Osinski, G.R. & Roehm, C.L. (2008) *Thermokarst lakes and ponds on Mars in the very recent (late Amazonian) past*. EPSL. 272, 382–393

Stahler, S.W., Shu, F.H. & Taam, R.E. (1980) *The evolution of protostars. I. Global formulation and results*. Astrophys. J. 241, 637–654.

Susko, D., Karunatillake, S., Kodikara, G., Skok, J.R., Wray, J., Heldmann, J., Cousin, A. & Judice, T. (2017) *A record of igneous evolution in Elysium, a major Martian volcanic province*. Sci. Rep. 7, 43177.

Tanaka, K.L. (2000) *Dust and Ice Deposition in the Martian Geologic Record*. Icarus. 144, 254–266.

Tanaka, K.L. (2001) *Geologic History of the Polar Regions of Mars Based on Mars Global Surveyor Data. I. Noachian and Hesperian Periods*. Icarus. 154, 3–21.

Tanaka, K.L., Robbins, S.J., Fortezzo, C.M., Skinner, J.A. & Hare, T.M. (2014) *The digital geologic map of Mars: Chronostratigraphic ages, topographic and crater morphologic characteristics, and updated resurfacing history*. Planet. Space Sci. 95, 11–24.

Tanaka, K.L., Skinner, J.A., Jr., Dohm, J.M., Irwin, R.P., III, Kolb, E.J., Fortezzo, C.M., Platz, T., Michael, G.G. and Hare, T.M. (2014) *Geologic map of Mars*. U.S. Geological Survey Scientific Investigations Map 3292, scale 1:20,000,000, pamphlet 43 p.

Thomas, N., Markiewicz, W.J., Sablotny, R.M., Wuttke, M.W. & Keller, H.U. (1999) *The color of the Martian sky and its influence on the illumination of the Martian surface*. J. Geophys. Res. 104(E4), 8795–8808.

Tosi, N. & Padovan, S. (2021) Mercury, Moon, Mars: Surface expressions of mantle convection and interior evolution of stagnant-lid bodies. In: Marquardt, H., Ballmer, M., Cottar, S. & Konter, J. (eds.). *Mantle Convection and Surface Expressions*. AGU Monograph Series.

Touma, J. & Wisdom, J. (1993) *The Chaotic Obliquity of Mars*. Science. 259, 1294–1297.

Treiman, A.H., Gleason, J.D. & Bogard, D.D. (2000) *The SNC meteorites are from Mars*. Planet. Space Sci. 48, 1213–30.

Trieloff, M. (2018) *The chronology of formation of solids and meteorite parent bodies in the early solar system*. EPSC Abstracts Vol. 12. EPSC2018-1205.

Vaucher, J., Baratoux, D., Mangold, N., Pinet, P., Kurita, K. & Grégoire, M. (2009) *The volcanic history of central Elysium Planitia: Implications for Martian magmatism*. Icarus. 204, 418–42.

Warner, N.H., Sowe, M., Gupta, S., Dumke, A. & Goddard, K. (2013) *Fill and spill of giant lakes in the eastern Valles Marineris region of Mars*. Geology. 41(6), 675–78.

Watters, T.R., McGovern, P.J. & Irwin, R.P., III (2007) *Hemispheres Apart: The Crustal Dichotomy on Mars*. Annu. Rev. Earth Planet. Sci. 35, 621–52.

Weiss, L.M., Marcy, G.W., Petigura, E.A., Fulton, B.J., Howard, A.W., Winn, J.N., Isaacson, H.T., Morton, T.D., Hirsch, L.A., Sinukoff, E.J., Cumming, A., Hebb, L. & Cargile P.A. (2017) *The California-Kepler Survey. V. Peas in a Pod: Planets in a Kepler Multi-planet System Are Similar in Size and Regularly Spaced.* Astrophys. J. 155, 48.

Wenzel, M.J., Manga, M. & Jellinek, A.M. (2004) *Tharsis as a consequence of Mars' dichotomy and layered mantle.* Geophys. Res. Lett. 31, L04702.

Werner, S.C. (2009) *The global Martian volcanic evolutionary history.* Icarus. 201, 44–68.

Werner, S.C. & Tanaka, K.L. (2011) *Redefinition of the crater-density and absolute-age boundaries for the chronostratigraphic system of Mars.* Icarus. 215, 603–7.

Wilson, L. & Head, J.W., III (2002) *Tharsis-radial graben systems as the surface manifestation of plume-related dike intrusion complexes: Models and implications.* J. Geophys. Res. 107(E8), 5057.

Wilson, R.J. (2011) *Dust cycle modeling with the GFDL Mars general circulation model.* Paper presented at the Fourth International Workshop on the Mars atmosphere, Paris, France.

Wilson, S.A., Grant, J.A., Howard, A.D. & Buczkowski, D.L. (2018) *The nature and origin of deposits in Uzboi Vallis on Mars.* J. Geophys. Res.: Planets. 123(7), 1842–62.

Wilson, S.A., Howard, A.D., Moore, J.M. & Grant, J.A. (2016) *A cold-wet middle-latitude environment on Mars during the Hesperian-Amazonian transition: Evidence from northern Arabia valleys and paleolakes.* J. Geophys. Res. Planets. 121, 1667–94.

Zhong, S. (2009) *Migration of Tharsis volcanism on Mars caused by differential rotation of the lithosphere.* Nat. Geosci. 2, 19–23.

Zimbelman, J.R., Garry, W.B., Bleacher, J.E. & Crown, D.A. (2015) Volcanism on Mars. In: Sigurdsson, H., Houghton, B., McNutt, S.R., Rymer, H. & Stix, J. (eds.). *The Encyclopedia of Volcanoes.* Academic Press. pp. 717–28.

Zimbelman, J.R. & Scheidt, S.P. (2012) *Hesperian Age for Western Medusae Fossae Formation, Mars.* Science. 336, 1683.

INDEX

A

accretion 31–2, 36
Acidalia Planitia 154
Airy-0 prime meridian 20, 21
alases 215, 216
Alba Fossae 234
Alba Mons 99, 160, 233–4
albedo 18
    of Phobos 85
Albor Tholus 175–6
ALH84001, Martian rock xiv, 123
altitude map 68, 69
Amazonian era 201–4
    climate 205–8, 209–13,
        215–19, 221–5
    dust cycle 227–31, 235
    equatorial glaciers 209–13
    high-latitude ice 215–19
    polar regions 221–5
    volcanism 233–6
Amazonis Planitia 170, 171
amino acids 125
Amphitrites Patera 98
Apollinaris Mons 182, 235
Apollinaris Tholus 176
Apollo missions 65
aquifers 90, 154, 165
Arabia Terra 230
Arcadia Planitia 154
Argentea Planum 193
Argyre crater 103, 132–3, 194
Arsia Mons 159, 160, 189
Ascraeus Mons 159, 160, 189,
    235
ash clouds 103

asteroid belt 34, 38, 39
asteroids 201
Athabasca Valles 235
atmosphere 15, 65, 79, 87–90,
        101, 103, 104, 114, 116,
        117, 119, 121, 132, 143,
        144, 180–1, 204, 221, 222,
        228, 241, 243, 244
    carbon dioxide component 87,
        88, 90, 103, 105, 119, 132,
        143, 244, 249
    sulphur dioxide component
        90, 143, 144
    water component 87, 88, 89,
        90, 103, 104, 119, 143
aureole, Olympus Mons 171–4
Aurorae Chaos 162, 165, 165–7
axis of rotation 107, 108, 120,
    185–9

B

bacteria, spread of human 246
Beer, Wilhelm 20
Biblis Tholus 160
black rock 104
black smokers 128, 137
bouncing barrier 31

C

calderas 158, 169, 176, 234
cancer risks 243
Candor Chasma 161, 162
Capri Chasma 162

carbon-based amino acids 125
carbon dioxide 15, 44, 87, 88, 90,
        103, 105, 116, 119, 132,
        143, 192, 209, 222, 223, 244
catena 19
cavi 19
Ceraunius Fossae 234
Ceraunius Tholus 160
Cerberus region 176
Ceres 187
chaoses 19, 151–5, 163, 165, 166–7
    Aurorae Chaos 162, 165,
        165–7
Chasma Australe 223
Chasma Boreale 223
chasmata 20, 161–7, 223, 224
    Candor Chasma 161, 162
    Capri Chasma 162
    Chasma Australe 223
    Chasma Boreale 223
    Coprates Chasma 162
    Echus Chasma 161
    Eos Chasmata 162
    Ganges Chasma 162
    Hebes Chasma 161
    Ius Chasma 162
    Melas Chasma 162
    Ophir Chasma 161
chondritic meteorites 64
chondrules 28, 31, 32, 32–3, 64
Chryse crater 103, 132
Chryse Planitia 154, 163, 166
Cimmeria plateau 135
Claritas ridge 132
clays 115, 117, 136, 138, 143
climate, Amazonian era 205–8,
        209
climate change, Hesperian era
        143–5, 191–4
Climate Orbiter satellite 240
coastline 121

cold sinks 192, 206
colles 19
    Scandia Colles 186, 192
colonisation of Mars 239, 247–50
comets 33, 88, 101, 107, 249
composition of Mars 47–51, 87
computer simulations 32, 39,
        74, 206
concentric crater fill 211
conduction heating 71, 72
convection currents 74
convection heating 71
Coprates Chasma 162
Coprates Mensa 166
core 48, 50, 71, 72, 73, 74–6, 180,
        204
crater counting 58–62, 68, 85,
        116, 158, 165, 182, 189, 201
craters (impact) 19, 53–5, 59–62,
        68, 87, 88, 104, 118, 148–9,
        186, 223
    Amazonian era 201–2, 208,
        211–13
    Argyre 103, 132–3, 194
    Chryse 103, 132
    Eridania 135–7
    Great Dichotomy impact
        theory 77–81, 97
    Gusev 135
    Hellas 21, 67, 68, 97–101,
        144, 228
    Korolev 207, 208, 212
    Martian moons 85, 86
    pedestal craters 182, 212–13
    Prometheus 223
    Stickney 85
crust, surface 50–1, 59, 68–9, 70,
        71, 72, 74–6, 79, 87, 90, 97,
        99, 103, 113, 114, 117–18,
        133, 137, 148, 151, 162,
        163–4, 180, 185, 187–8

cryosphere 147–9

D

dam bursts 19, 165–6
dating Martian features 58–62
day length 13
debris disc, protoplanetary 35–7, 99
Deimos 83, 85–6, 249
deposition 14, 87
DNA 126
dorsa 20, 192, 193, 221, 224
    Dorsa Argentea 192, 221, 224
drift barrier 32
dust cycle 227, 235
    duststones 230
    electrical charge/static 229, 230
    hazard to human visitors 243
    storms 228–9, 245

E

Earth 15, 34, 37, 38, 58, 59, 67, 127
    crust 73, 158, 187
    mantle 72–3, 158
eccentricity, orbit 107–8, 109
Echus Chasma 161
ejecta 64, 86, 97, 104–5, 139, 212, 213
ejecta blankets 53, 54, 67, 68, 79, 149, 212
electromagnetic dynamo, core 75, 75–6
Elysium 18, 175–8, 181, 207, 233, 248
Elysium Mons 175–7, 207, 228, 235

Elysium Planitia 235
Eos Chasmata 162
Eos Mensa 166
equator 20, 21, 105, 112, 114, 185, 187, 206, 211
Eridania 18, 135–9
eskers 192, 193, 221
exoplanets 33
extremophiles 126–7, 144

F

fault lines 98, 106, 113, 131, 234
fetch, wind 121
fissures 104, 113, 118, 132
flood basalts 104, 144, 163, 175, 176, 235
flooding 115, 149, 151, 153, 161, 163, 166–7
fluctus 19
fossae 20
    Alba Fossae 234
    Ceraunius Fossae 234
    Medusa Fossae Formation 179–83, 212, 230
    Tantalus Fossae 234
fossils, Earth 127
fossils, Martian 123–4
'free air' gravity 70

G

Ganges Chasma 162
Gemina Lingula 223
geysers 49, 103, 128, 148
glaciers 120, 144, 149, 202, 203, 207, 209–11
global chilling 144–5
graben 131, 132, 188
Grand Tack theory 39, 100, 101

gravity 5, 13, 31, 32, 36, 37, 38, 70, 83–4, 86, 89, 93, 108, 137, 228
gravity maps 21, 69–70, 185
Great Dichotomy 68–70, 90, 97, 105, 108, 111, 117, 163, 164, 176, 185, 189, 206
   boundary 112–13, 131
   convection theory 71–6, 80
   impact theory 77–81, 97
greenhouse effect 88
Greenwich Meridian 20
groundwater 103, 105, 151, 153, 163, 164
Gusev crater 135

H

Hadriaca Patera 98
haematite 180, 181, 183
half-lives, radioactive 48
Hawaii, volcanic activity 173, 189
Hebes Chasma 161
Hecates Tholus 175–6
helium 35
Hellas crater 21, 67, 68, 97–101, 144, 206, 207, 228
Hesperian era
   climate change 143–5, 191–4
   cryosphere 147–9
   Elysium 175–8
   ice caps and glaciers 191–4, 210
   Medusa Fossae Formation 179–83
   Olympus Mons 169–74
   polar shift 185–9, 191–2
   valles and chaoses 151–5
   Valles Marineris 161–7
   volcanoes 157–60, 169–74, 175–8

hot Jupiters 38
human mission to Mars 242–6
hydrogen 26, 35, 44
hydrothermal vents 106, 128, 137, 148

I

icebergs 120
ice on Mars 20, 115, 116, 138, 147–9, 151, 153, 244
   Amazonian era 202, 203, 206–7, 209–13
   equatorial 209–13
   glaciers 120, 144, 149, 202, 203, 206, 207, 209–11
   high-latitude ice 215–19
   ice caps 116, 145, 152, 186, 191–4, 206, 209, 218, 219, 221–5, 227, 249
   ice melt 87, 186
   see also polar regions; water on Mars
impact melt 53, 54
infrared light 71
insulae 19, 166
International Astronomical Union 18
interstellar/giant molecular cloud 25–9
iron-nickel-sulphur core 51, 71, 74
iron oxide in rocks 123
Ius Chasma 162
Ius Mensa 166

J

Jupiter xiii, 34, 37–9, 47, 83, 100, 108

K

Korolev crater 207, 208, 212

L

labyrinthus 20
   Noctis Labyrinthus 162
lacus 19
Ladon river 132
Lagrange points 78
lakes/lake systems 144, 145, 161,
   165, 193–4, 203
   Lake Eridania 135–9
landers 20, 64, 65, 166, 217, 218,
   240–2
landslides 171–2, 202
lasers, space-borne 63
Late Heavy Bombardment
   99–101
lava 79, 87, 100, 104, 138, 149,
   157, 158, 159, 162, 163,
   164, 167, 169, 170, 171,
   177–8, 203, 234, 235
   flows 57–8, 60, 68, 113, 133,
     138, 158–9, 163, 170, 171,
     172
law of superposition 57–8
life, creation of 124–6
life on Mars 123–4, 125–9, 144,
   148, 245, 246
lightning 229
light radiation 71
linear tension faults 234
lithium atoms 35
lunar basins 100

M

Ma'adim Vallis 135, 138
Mädler, Johann 17, 20

magma 75, 98, 113, 118, 131,
   143, 157, 158, 163–4, 176,
   177
magnetic fields 75, 89, 132
magnetite 75, 180
magnetometers 63, 240
magnetotactic bacteria 123
Malea Patera 98
Malea Planum 186
mantle 51, 58–9, 70, 71, 72, 73,
   73–6, 80, 97, 98, 117, 148,
   151, 170, 180, 187–9
   convection 72, 73, 80, 117
   plumes 73–4, 75–6, 112–13,
     117, 131, 133, 163, 175,
     176, 189
mantle-crust boundary 72–3,
   74–6, 104
mantle, Earth's 72–3
map quadrant names 18
maps of Mars 17–21, 68, 69–70
Mariner 9 space probe 17, 20
Mars datum 21
Mars Pathfinder 166
marsquakes 99–100, 152, 202
mass extinctions 127, 128
mass of Mars 34, 47, 108, 185
Mauna Kea, Hawaii 169
measuring Martian relief 21
Medusa Fossae Formation
   179–83, 212, 230
Melas Chasma 162
melting event 49, 87, 117, 132,
   180
mensa 19
   Coprates Mensa 166
   Eos Mensa 166
   Ius Mensa 166
Mercury 34, 37, 38, 47, 78, 83,
   107
Meridiani Planum 242

meteorites xiii, 28, 54–5, 59–62, 64–5, 77–9, 87, 88, 89, 101, 104, 117, 121, 127, 139, 202, 212
  hitting ocean 121–2, 145, 149
  Martian xiv, 64–5, 123
methane 129, 244
Miller, Stanley 125
mineral-rich water 117–18, 128, 137
molten rock 49, 53, 54, 121, 149, 163, 177
mons 19
  Alba Mons 99, 160, 233–4
  Apollinaris Mons 176, 182, 235
  Arsia Mons 159, 160, 189
  Ascraeus Mons 159, 160, 189, 235
  Elysium Mons 175–7, 207, 228, 235
  Olympus Mons 21, 157, 160, 169–74, 177, 211, 228
  Pavonis Mons 159, 160, 187, 189
  Uranius Mons 160
montes 19
  Phlegra Montes 207
  Tharsis Montes 170, 187, 188–9, 207, 211, 228, 234
moons, Martian 7, 15, 63, 83–6, 108, 120
Moon, the 15, 61, 62, 65, 67, 78, 99
Morava river 132

N
Neptune 34, 39, 100, 101
Nix Olympica 169

Noachis Terra 67
Nochian era 67, 68, 81, 90, 97, 189
  early 103–6
  Lake Eridania 135–9
  northern ocean 104, 106, 119–22, 128
  potential for life 123–4, 125–9
  Tharsis Plateau 111–14, 117, 120, 131–3, 163, 164
  water 115–18, 119–22, 132–3, 135–9
Noctis Labyrinthus 162
northern hemisphere 68, 70, 73, 75, 76, 79, 80, 81, 90, 103, 107, 113, 114, 119–22, 131, 163, 164, 185–6, 206, 209, 212, 215–19, 227, 228
  see also Great Dichotomy; ice caps; polar regions
north polar basal unit 224
nuclear fusion 35

O
obliquity, planetary 108–9, 145, 185, 191, 203, 205, 206, 207, 210, 218, 221, 225, 249
ocean, northern 104, 105, 119–22, 128, 135, 143, 145, 148, 174, 186
Olympia Planum 224
Olympus Mons 21, 157, 160, 169–74, 177, 207, 211, 228
Ophir Chasma 161
orbital resonance 38, 39
orbit of the Sun, Martian 15, 107–9, 205
oxygen 28, 44, 94, 119, 179, 180, 183, 244

P

palus 19
paterae 19, 98
Pavonis Mons 159, 160, 187,
        189
pedestal craters 182, 212–13
Peneus Patera 98
Perchlorates 243, 245
periglacial landforms 215–16
permafrost 144, 147–8, 151,
        152, 153, 154, 202, 203,
        209, 215
Perseverance rover 63–4
Phlegra Montes 207
Phobos 83–6, 249
Phoenix lander 217–19
photography 63, 240
pingos 215, 216
Pityusa Patera 99
plana 19
    see also plateaus
planetary embryos 35–9, 47,
        78, 99
planetesimals 32–3, 36–9, 61, 64,
        87, 99, 100
planitia 19
    Acidalia Planitia 154
    Amazonis Planitia 170, 171
    Arcadia Planitia 154
    Chryse Planitia 154, 163, 166
    Elysium Planitia 235
    Utopia Planitia 97, 154
plateaus 19, 212
    Arabia Terra 230
    Cimmeria plateau 135
    Noachis Terra 67
    Sirenum plateau 135
    Syria Planum plateau 131
    Tharsis Plateau 111–14,
        117, 120, 131–3, 144, 157,
        159–60, 175, 176, 177, 228

plate tectonics 58–9, 73, 176, 233
polar axis 185–9, 191–2
    see also obliquity
polar regions 17, 83, 108, 206,
        209, 215–19, 221–5, 227,
        249
polar wander 185–9, 191–2
polygonal cracks 215, 216
prime meridian 20
probes 9, 17, 63, 239, 240
Proctor, Richard 17
Prometheus crater 223
proto-Earth 78
proto-Mars 37, 38, 39
protoplanetary disc 27–8, 31–4,
        35
proto-Sun 35–6, 87
pyroclastic material 177, 178,
        182

R

radiation/radioactivity 26, 48, 49,
        71, 243, 245
radiometric data 61, 63, 65
rain on Mars 44, 94, 95, 104, 105,
        116, 118, 119, 137, 143,
        144, 145, 158, 186
red colouring 179–83
rift valleys 114, 162
rivers/river systems 105, 132,
        135–6, 137, 147, 164, 185,
        193–4, 203, 216, 217
    riverbeds 105, 115, 121, 193
robots 9, 239
rock glaciers 153, 210–11, 213
rotation, planet 75
rovers 4–6, 63–4, 115, 135, 166,
        245
rupes 19
rust 179

S

saltation 183
satellites, man-made 69, 76, 115, 116, 118, 240
satellites, Martian 5, 21, 69
Saturn xiii, 34, 37, 38, 83, 100
Scandia Colles 186, 192
Schiaparelli, Giovanni 17, 20
Schiaparelli lander 242
scopuli 19
seabed 105, 122, 145
seas 104, 135, 145, 173
    northern ocean 173
seasons 107–9, 115–16, 191, 205, 206, 227
    see also summer; winter
secondary crust 51
shield volcanoes 169–74, 176, 189, 234
Shklovsky, Iosif Samuilovich 84–5
shock-glass 65
shorelines 115, 120, 138
sinkholes 118, 147
sinus 19
    Sinus Meridiani 20
Sirenum plateau 135
size of Mars 15, 49
sky colour 105, 180, 181
snow 28, 32, 144, 148, 191, 218
solar system arrangement 33–4, 36–9
solar system, formation of 27–9
solar tides 120
solar wind 89, 117, 132, 145
sols 13, 86
southern hemisphere 68, 70, 73, 75, 76, 80, 81, 90, 103, 104, 113, 114, 135–9, 158, 163, 164, 185–6, 206, 209, 212, 219, 227, 228

    see also Great Dichotomy; ice caps; polar regions
Spirit rover 135
steam 44, 87, 88, 115, 122, 132, 138, 149
Stickney crater 85
storms 105, 119, 228–9
sulphates 143
sulphides 103
sulphur 104, 132, 145
sulphur dioxide in Mars's atmosphere 90, 144
summer 107, 108, 109, 115–16, 205, 217, 218, 228
Sun, the 3, 7, 15, 27, 35, 89, 107, 108, 117, 205, 206, 218, 227, 228, 229, 243
supernovas 26
superposition, law of 57–8
Syria Planum plateau 131

T

Tantalus Fossae 234
terrae 19
    see also plateaus
terraforming Mars 249
tesserae 20, 215
Tharsis Plateau 111–14, 117, 120, 131–3, 144, 157, 159–60, 175, 176, 177, 228
    Amazonian era 228, 233–4
    polar wander 187–8, 189
    Tharsis Montes 170, 187, 188–9, 207, 211, 228, 234
    Valles Marineris 161–7
Tharsis Tholus 160
Thaumasia Planum 228
tholus 19
    Albor Tholus 175–6
    Apollinaris Tholus 176

Biblis Tholus 160
Ceraunius Tholus 160
Hecates Tholus 175–6
Ulysses Tholus 160
Zephyria Tholus 176
tsunamis 104, 120, 121–2, 149,
    166–7, 174
Tyrrhena Patera 98

U
ultra-violet light 90
Ulysses Tholus 160
undae 20
universe, formation of the 25–6
uranium 48
Uranius Mons 160
Uranus 34, 39, 100
Urey, Harold 125
Utopia Planitia 97, 154
Uzboi river 132
Uzboi Vallis 132

V
valles 19, 105, 131, 151, 154, 155
    Ma'adim Vallis 135, 138
    Uzboi Vallis 132
    Valles Marineris 161–7
Valles Marineris 161–7
vastitas 19
    see also northern hemisphere
Vastitas Borealis 154, 186, 207,
    217
vents 44
Venus xiii, 34, 37, 38, 83, 187
Viking 1 lander 21, 65, 166
volcanoes/volcanism 3, 19, 73,
    99, 103, 104–5, 111–14,
    118, 133, 143–5, 147, 149,
    157–60, 163, 169–74, 203

Amazonian era 228, 233–6
Elysium 175–8, 207, 228, 233
Olympus Mons 21, 157, 160,
    169–74, 177, 211, 228
    see also Tharsis Plateau

W
water on Mars 66, 87–8, 90,
    104–6, 109, 115–18,
    119–22, 126, 132, 135–9,
    144, 145, 148, 149, 151–4,
    163, 186, 206, 209, 216–17,
    218, 221, 235
    groundwater 103, 105, 151,
        153, 163, 164, 216
    see also ice on Mars
waves 121, 137
weather 7, 8, 20, 63, 105, 119, 138
weather stations 6, 63
weight of Mars 15
wet-based glaciers 210
winds 20, 116, 121, 148, 180,
    182–3, 202, 206, 213, 223,
    227, 228, 228–9
winter 107, 108, 115–16, 144,
    205, 218, 222

X
X-rays 90

Y
yardangs 182
year length, Martian 13

Z
Zephyria Tholus 176
zero level – Mars datum 21

ABOUT THE AUTHOR

Dr Simon Morden trained as a planetary geologist and geophysicist, realised he was never going to get into space, and decided to write about it instead. His award-winning writing career blends narrative science, science fiction, fantasy and horror. He is a past winner of the Philip K. Dick Memorial Award for his Metrozone series of novels set in post-apocalyptic London. He lives in Gateshead, Tyne and Wear, with his wife and family.